Service Management in Computing and Telecommunications

For a complete listing of the *Artech House Telecommunications* Library,
turn to the back of this book.

Service Management in Computing and Telecommunications

Richard Hallows

Artech House
Boston • London

Library of Congress Cataloging-in-Publication Data
Hallows, Richard
Service management in computing and telecommunications / Richard Hallows.
Includes bibliographical references and index.
ISBN 0-89006-676-0
1. Telecommunication—Customer services. 2. Computer industry—Customer services.
I. Title.

HE7661.H35 1995 94-36072
004'.0068'8—dc20 CIP

British Library Cataloguing in Publication Data
Hallows, Richard
Service Management in Computing and Telecommunications
I. Title
621.382068

ISBN 0-89006-676-0

685 Canton Street
Norwood, MA 02062

International Standard Book Number: 0-89006-676-0
Library of Congress Catalog Card Number: 94-36072

10 9 8 7 6 5 4 3 2 1

Contents

Preface

This book is the result of ten years' personal experience working in various aspects of the provision of public services with a variety of service providers. I am not now, nor have I ever been, a consultant of any kind. Everything that has been learned has been learned the hard way—by doing it, or at least being around while it was being done. Service management has grown in importance over that time, and this will only accelerate in the future for a variety of reasons.

Technical developments have changed the nature of the problems associated with service provision over the years. It has taken away some of the emphasis on systems management and placed increased importance on the management of the service, particularly the management of the users of the service. The computing power available is in some ways effectively infinite, as is digital storage, and if the proponents of asynchronous transfer mode (ATM) developments are to be believed, so is network capacity in the future. Technical constraints on service capabilities are rapidly disappearing.

In the public arena this means that the basis of competition will change as the networks and products being offered become increasingly similar in terms of feature and function. The competitive battleground will be the management of the service and the management of the customer. This is equally important in the provision of private services internal to an organization, as the private service provider is not going to be allowed to avoid the impacts of competition. Outsourcing of data processing and networking capabilities to public operators is increasing in frequency and scale. This potentially places the internal corporate information technology and telecommunications managers in competition with public service providers.

The management of the service has a direct impact on success and revenue generation in the public environment. A well-managed service should encourage increased usage and should generate repeat business for the provider as well as expand existing business. Management in the private service environment does the same. It is true that service providers get the customers they deserve, in terms of both number of customers and type. In fact it is possible to go even further and state that service providers get the customers they create. Service management is a technique for service providers to create the type of customer they want to use the service.

Equally it can be said that the users of the service get the service provider they deserve, and the service provider they can help to create. Service management is not necessarily a one-sided function, and the development of a partnership between the services provider and the user community is an essential element of ensuring the provision of quality services. Service management is important to everyone, whether they provide or receive services, good, bad, or indifferent.

Acknowledgments

There are a lot of people who wittingly or unwittingly have helped me with this book. I would like to thank everyone at Cable & Wireless Business Networks, especially Duncan Lewis and Rob Fisher. I would especially like to thank three people who I have worked with over the years, Clynt Stubbs, Brian Rastrick, and Bob Rutherford, who, as a consequence of working with me, all have gray hair.

Acknowledgments

1 Introduction

Early in 1992 I was invited to attend a meeting with a public telecommunications operator. Although that meeting had little or nothing to do with the provision of customer service, it was, by accident, responsible for the writing of this book.

The meeting room still contained the debris from a previous meeting, including a set of notes written on a board on the wall. In large letters at the top of the board was the question the previous meeting had been trying to address:

"Why don't our customers like dealing with us??????"

The way it was written, in black capitals with an inordinate number of question marks, made it look like a plaintive cry for help. I suspect it had been a highly emotional meeting.

The fact that the question was being asked meant that the organization had already made two large steps forward in its provision of service to its customers. First, it had acknowledged that its customers didn't find doing business with them an enjoyable experience; second, it had recognized that this was important, not only to the customers, but to the company itself. What is perhaps even more remarkable is that the company in question is, at this time, a monopoly supplier.

Beneath the question on the board were listed some of the reasons customers had cited for the difficulty they had dealing with this supplier:

1. There are too many numbers to call.
2. I never get through to the person who can help me.
3. I never know what is going on to resolve my problems.
4. I get different answers from different people.
5. It takes too long to resolve a problem.
6. Nobody asks me what I want.
7. I don't know what's supposed to happen.

Either the meeting had gotten too depressed to carry on or had run out of time, because the list ended with

8. And a lot more!!!!!!

It is an interesting list. The company clearly felt it had been providing a level of service to its customers. There were people in place to help their customers in any number of situations. There were numbers to call if a customer ever needed help, and there were support staff who could help to deal with more complex problems that could not be resolved over the telephone. Clearly there were procedures in place to ensure that the calls were passed to the right place within the organization. The problems were not those associated with lack of desire to provide a service to the customer (as the plaintive nature of the question shows), but of not managing the provision of the service to meet the customer's needs.

The one thing that hadn't been identified was why the company was spending the time to answer these questions; after all, they were a monopoly and making healthy profits the way they had always worked in the past, and the quality and range of their products was excellent. The answer was easy. The perceived that lack of service to their customers was costing them money. Customers were looking for alternatives, the monopoly status was being removed in some areas of the business, and competitive suppliers were making inroads into the customer base, and at the same time that they were losing customers for not providing an acceptable level of customer service they were spending more on "customer support" than ever before.

The company had seen the direct impact managing service provision has on both the cost and revenue lines of the organization. This is increasingly being acknowledged in service-oriented organizations, with some even making achievement of a certain level of customer satisfaction the sole determination of pay bonuses.

The level of service provided—and the management of service capabilities to achieve set levels of service performance—has become a major competitive differentiator. Profitability in the service industry is heavily affected by the length of the relationship between customer and supplier. The longer the service relationship, the more profitable it is. Service management is designed to encourage longevity.

Service management encompasses every area of concern to the customer of communications services, from the management of physical infrastructure to the management of the way in which an employee of the service provider treats the customer on a personal level. It encompasses every element of the service provider organization with one measure by which they can be judged: the satisfaction of the customer.

As with many problems, understanding that it exists means that an organization has already made significant progress in resolving it. However, having taken the steps to understand that the problem exists, the organization now finds itself in uncharted waters. This book aims to provide the navigational tools to enable an organization to complete the journey to satisfied customers.

The scope of what is needed to successfully manage the provision of service cannot be underestimated, but it can be broken into discrete elements, each of which will need to be addressed over time. As one element of the changes required is cultural, it is likely to be a long process, but one that is measurable at each stage by improved customer and employee satisfaction.

Before progressing into uncharted waters it is useful to know where we are starting from. In asking the question "Why don't our customers like dealing with us?" the starting point is being defined. This is one of the most important elements of any implementation of service management principles. There are few "green field" opportunities that exist for the implementation of a service management culture and organization, and any implementation will have to work from a base of the existing infrastructure and culture. Defining the starting point from which service management can be implemented helps to recognize the extent of the problem. Just as importantly, it should also help to determine the existing strengths from which customer satisfaction can be enhanced.

Once you know where you are, it is worth having a clear picture of where you would like to be. It is unlikely that any service management journey will ever actually end; it is more likely to be a series of stages that need to be reached along the way. It is very difficult to define an end point. This is partly because any end point that exists is not defined by where you want to be, but more by where your customers want you to be. As customers needs and objectives change, they in turn change the objectives and scope of the service management implementation.

This indicates two things: first, that it is essential for the supplier to completely understand what drives customer needs, and second, that the service business is not for the faint hearted. Both these themes will recur throughout the book.

There are many ways in which the determination of customer need can be approached. The most obvious way is to ask the customer. However, it may be that the implementation of a plan for service management to meet the needs articulated in a customer "wish list" is not the best way forward. It must be acknowledged that although customers are the final arbiter, they do not always know what they want, and it can be more productive to work from defining what the customer does not want.

It is probably not appropriate to start from a position that customers do not know what they want. However, it is often the case that the customer has had unrealistic expectations as to what a service can and cannot provide. They may have read hype in a magazine or been oversold by a salesperson. The setting of expectations is a very important precursor to the provision of services. It is often looked on as not being a part of the service, but working with the customers to determine their requirements is an essential part of the service that can be provided.

The service management journey is one that must be taken by both the service recipient and the service provider. At the same time as the service being provided develops, the service required will also develop—and probably at a faster pace. An additional challenge of any service management implementation is to manage the changing needs and expectations of customers.

It is essential to ensure that the customer is involved in the development of service management by the service provider. Although there are dangers associated with the changing needs of the customer, the partnership must be maintained despite these tensions. When all the parties involved in service provision are included in the development of the service, the level of acceptance of the outcome is much greater. When the customers have a say in what is provided, they are much more satisfied with it.

The management of the partnership with the customer and the management of the development itself is a significant factor in determining the satisfaction of the customers. The management of the customer itself, through the generation of the correct expectation can determine the eventual perception of the service. If the customer expectation is set too high, then although a good quality service may be provided, it may still be perceived as being poor if the service does not match that expectation. The management of expectation and perception are two sides of the same coin. Both have to be managed to determine the eventual levels of customer satisfaction.

There are many tools available for the management of expectation, the most frequently used being the service level agreement. This is formal documentation of the quality of service the service provider is willing to commit to the customer. However, the service level agreement has normally been interpreted by the service provider as an encumbrance forced upon them by the customer.

Customers do view the service level agreement as a mechanism to manage the supplier, but it should work both ways. A service level agreement is a very effective tool that can be used by the supplier to manage the customer within the bounds of the service levels defined.

All of the elements of service management that need to be addressed have a serious impact on the whole of the service provider organization. Although the initial focus of any implementation of service management programs may be on the front line staff who deal directly with the customer, they are dependent on the whole organization behind them for support. Customers of a public communication service provider with whom I have come into contact are increasingly asking to see how the organization is structured and managed in order to see whether that support is easily available to the front line staff they deal with on a daily basis. In order to ensure that a customer focus is achieved and that the right support is available, a service culture must exist throughout the organization, and management systems need to be implemented in areas not traditionally associated with the provision of customer service to ensure that service is provided. The inability of a personnel department to recruit the right staff quickly enough to provide customer support functions has a direct impact on the quality of service provided to the customer. Everyone in a service provider organization is responsible for customer service.

In trying to define the bounds of a service management implementation, we need to have a clear view of where we are and where we are trying to get to. As we progress along the journey it will be important to continually check that we know where we are at any point in time. The measurement and monitoring of the quality of service provided underpins much of what else needs to be achieved. Managing expectations, perceptions, and customer satisfaction depends upon being able to monitor and measure all of the aspects of service provision perceived by the customer.

Judging from the problems articulated on the board in the meeting room, many of the elements of service management are missing from the way in which the service being provided to those customers is being managed. Where the customer has an expectation, the perception is that it is not being met. In other areas the customer does not even have an expectation defined. Managing quality and customer satisfaction without clear management of expectation and perception is like starting a journey without knowing where you are or where you want to go.

Not all of the elements of service discussed in this book may be necessary in all business environments. The key is to provide the level of service that is appropriate to the customer. There is no single answer for any one organization, but the questions are normally the same. The key question for all service management implementations is whether what is being done will create a measurable improvement in customer satisfaction. If this is kept in mind at all times it should ensure an effective and worthwhile implementation.

What Service Is Being Provided? 2

Before setting out on any journey it is useful to know where you are starting from. In the development of a service management implementation there are several reference points that must be taken into account in order to define the starting position. A clear picture of the starting point determines the scope of the task ahead. It will normally also provide a clear picture of where the most benefit can be gained early in the implementation. This is important because it is vital that the implementation (which will undoubtedly be hard work) show real benefit early on. Tangible early examples of the value of service management will help to establish credibility and will smooth the later stages of the implementation.

2.1 REVIEW OF CURRENT SERVICE

The first step is a review of all aspects of the current service that is actually being provided and how it is being managed. This should provide a base line of the service that is in place today.

2.1.1 Who Are the Customers and Users of the Service?

Part of the service review is to try to identify and understand who is using the service. This is important as the features and functions of the service required are very different for different customer or user types. If, for example, a very technically aware and skilled user base exists, then service requirements for help desks and training may be minimized, whereas with a low skill level user base, these requirements may be paramount.

Understanding the users and their needs is fundamental to ensuring the value of a service management implementation. It is also the reason why it is not possible to define a service management check list of what should be done and when. Every implementation will be different, as it will be based on different users with different needs.

2.1.2 What Are the Users Doing with the Service?

Although sometimes very difficult to identify, it is important to know what the service is being used for, as this again will act as a determining factor of the functions and features of the service. For example, an electronic mail service designed for the transfer of electronic notes and documents would not be an adequate vehicle for the distribution of large manufacturing designs or the distribution of files containing hundreds of megabytes of software modules. This example may sound slightly ridiculous, but it has actually happened. A service I was involved with provided electronic mail capabilities to a range of users for the distribution of small notes and documents. What some of the users needed was a more generalized file transfer service to send software updates to their branch offices. The users had never been made aware of exactly what the service was intended to provide, and we didn't know what they wanted to use it for. Both of us were extremely unhappy with the outcome. The customer had a service that didn't meet their needs, and the service provider had a customer who caused constant problems for the service and for other users.

2.1.3 What Service Performance Parameters Are Being Measured?

Before any improvement in service performance can be implemented you need to know what service performance is currently being achieved. In order to know what is being achieved it is necessary to measure it and report against it. Without the measurement and reporting to enable an understanding of current service performance it is not possible to identify areas where improvements can be made.

Measurement and reporting are often looked on as "nice-to-haves" and not as essential parts of the provision of service to the customer. I suspect that this is in part due to the fact that the systems and processes that need to be implemented to measure and report service are perceived as indescribably dull and bureaucratic. Service providers often feel they know intuitively what the problems are. This can, however, be a one-sided view. Measurement of the elements of service perceived by the customer rather than the service provider should provide a better understanding of the service elements that need attention. Intuition is not enough to ensure the provision of quality services.

2.1.4 What Service Performance Is Being Achieved?

On the basis of the performance parameters that are being measured, the current levels of performance being achieved can be identified. In addition to identifying the current level of performance it is valuable to be able to obtain historical measurements and identify whether there are any trends in the levels of performance being achieved. This is important because the direction of the trend, rather than any absolute value, can set the levels of expectation and satisfaction in the mind of the users.

2.1.5 What Are the Key Processes in Place to Manage the Service?

The service management processes, while no guarantee of success, can be the key to preventing failure. An identification of the documented processes in place—ranging from change and fault management processes to the service development process and customer satisfaction management—provides a clear indication of the way in which service is delivered. It is the processes that normally determine the way in which the service organization interacts with the customers. This interaction goes a long way in determining the perception of the service among the user community. It is the tangible element of service that they see.

2.1.6 What Are the Key Groups Involved in the Delivery of the Service?

In addition to identifying the users of the service and the processes by which the service is delivered to these users, it is important to have a clear view of which groups are involved in the delivery of service to the users. Service delivery staff will normally find the most efficient way through an organization to help them deal with a customer problem. They often bypass all formal routes and rely on personal contacts and informal networks of people. Often the best way to get a problem resolved does not involve any of the known support groups. If this is the support process that works to deliver service, then it should be acknowledged and incorporated within the bounds of the management of the service.

This approach has become more prevalent in recent years with the identification of "virtual teams" and a less hierarchical approach to service delivery organizations. Within the support channels it is essential to know who is involved. Providing service support day-in, day-out requires everyone to understand the role they play in that support and whom they can call on for help. One of the dangers of the informal support channels is that they can provide exceptional service, but erratically. By incorporating informal channels within the service management implementation it may be possible to retain the exceptional service, but provide it consistently.

When the above identifications have been made, a clear overview will emerge of the service that is being delivered. It should show what levels of performance are being achieved, who is receiving and using the service, and who is providing the service and how.

2.2 REVIEW OF SERVICE SPECIFICATION

Surprisingly, it is often the case that services are being provided without any clear definition of the service that is *supposed* to be delivered. If you are lucky there will be a document with the title of "Service Specification" or "Service Requirements." If you are not lucky then the only reference point as to the service definition will be in the head of someone who has long since moved on to a different function.

The likelihood of the service currently being provided matching the service specification (should one exist) is probably very small. Changes to services are implemented in response to the day-to-day demands of users, normally without looking back at the original definition of what is to be provided. Yet this definition should provide a basic reference point for the implementation and management of the service.

The reason why this element of the starting point is important is because without knowing what the service is, it is difficult to determine the causes of the success or failure of the service. This is probably shown most clearly in the public service environment where poor perception of a service can be created by a wide variety of causes. The service could be meeting every requirement that has ever been defined but still have severe user satisfaction problems. It could have been oversold by an enthusiastic sales person, thereby setting customer expectations the service was never designed to meet. The requirements defined could have failed to reflect what the users really wanted. User requirements could have changed over time, making the original requirement definition obsolete, even if it was right at the time.

These possible reasons for user dissatisfaction show that it is not necessarily the day-to-day service provision that is the cause of service problems. Service management has to go much deeper than just the delivery of service to ensure that all aspects of service provision work in harmony to achieve customer satisfaction. It must also include service definition and service implementation. The service definition is an essential component of the service that enables service management to be successful.

2.3 REVIEW OF USERS' SERVICE EXPECTATIONS

This reference point refers directly to the users of the service. The expectations of the users is something that is constantly changing. Expectation is determined by such a wide variety of factors, ranging from their current experience of the service, to the way in which the service was sold to them, to the most recent article they have read about advances in new technology. The expectation of the user will determine their level of satisfaction with the service that is being delivered. With careful management of expectations, the service that is being delivered to them can achieve higher levels of user satisfaction. Service that is better than expected can increase customer satisfaction, whatever the real level of service being provided.

2.4 REVIEW OF USERS' PERCEPTION OF THE SERVICE

Like the review of service expectations, this reference point refers directly to the users. In some ways the users' perception of the service is the best measure of the service that is being delivered, as it combines both the measure of the service performance and the expectation of the user community.

The procedure for identifying the perceptions of the users needs to be structured to ensure that the best information possible is received. It should aim to discover as much about the user community as it does about the service being provided. In particular, understanding the perceptions of the customer provides a clear picture of the elements of the service that are most important to the user, and therefore most significant in determining perceptions.

The four starting points for any service management implementation—the service being delivered, the service that has been defined, the user expectation of the service, and the user perception of the service—may be different for any particular service management implementation, but they must all be understood. Sadly, it can probably be guaranteed that the four will bear almost no resemblance to one another. The objective of a service management implementation is to manage all the elements of the service, from the setting and ongoing management of the customers expectations to the delivery of service on a day-to-day basis to ensure that the gap between user expectation and user perception is narrowed to a point where the customers satisfaction with the service is at an acceptable level.

2.5 CONCLUSION

It is not possible to manage any one element of the service and achieve customer satisfaction without managing the other elements. Every aspect of the service interacts in such a way as to continually renew the challenges of service delivery. If, for example, the overall service delivery improves dramatically, but the elements of the service that are really important to the user do not change, then the user perception of service performance will remain poor, and the effect of the improvements on customer satisfaction will be negligible. If expectations are not managed, then the users may expect the improvements to continue beyond what has been achieved and again the effect on customer satisfaction will not be significant.

The integral nature of these four elements of the service management challenge also mean that it is not possible to undertake a "smoke and mirrors" exercise to improve customer perception without any real improvement in service quality, however tempting this may be when the problems of improving the service performance seem insurmountable. In fact, any such attempt will probably result in a reduction in customer perceptions of quality as it will raise expectations without raising the equivalent reality of the user experience.

The importance of the interrelationships also means that there are few quality absolutes that apply within the realm of service management. For example, a service availability achievement of 99.98% may sound impressive as an isolated absolute figure, but if the user expectations are set at 99.99%, then the users perception of the service is likely to be poor. If, however, expectations are 99.97% and delivery is 99.98%, then the perception is more likely to be good. Perception is improved by service that is better than expected.

Furthermore, the importance of perception and expectation also means that service management needs to be highly sensitive to service trends and the historical performance of the service. If the service is improving, then perception is likely to improve even if the service has yet to reach the levels of expectation. Sensitivity to trends, however, also encompasses trends that are outside the scope of the service; that is, *external* trends can also set expectations of the service. For example, say a user is a customer of a network service that is offering and delivering 99.97% availability, but the industry trend and the competitors reaction to that trend is to offer 99.98%. Even without any degradation in the service being offered, user satisfaction and perception of the service will often be adversely impacted.

The complexity of these relationships means that all the aspects of service management require constant attention both to themselves and to the way in which they interact with each other. The measurement of the success of service management is in itself very complex due to these interactions, and it is difficult to find any single way to measure both the service management success and the service quality. Again, it is probable that there are no absolute measures of success since elements will change every day, but again, it is trends in service achievement that measure the success of the management of the service.

Knowing where we are starting from sets the beginning of the trend and the starting point for managing the elements that determine the quality of service being delivered. The starting point for improving the levels of service and for identifying the best route to achieving customer satisfaction is to define what the service is actually required to do.

Defining and Understanding Customer Requirements 3

Any service environment implemented by a service provider should be based on a very clear understanding of the requirements that environment is trying to meet. This may sound obvious, but it continues to surprise me how many well-intentioned service initiatives are undertaken that nobody wants or needs, or how many are implemented without any real understanding of what they are trying to achieve.

Defining the requirements of the service is a fundamental starting point for implementing any service management system, but too often there are implicit requirements that are never fully evaluated or understood, and each of these has an impact on the overall perception of service that is created.

3.1 PROBLEMS IN DEFINING SERVICE REQUIREMENTS

There are three major problems with defining service requirements that every service provider organization needs to overcome in order to provide a stable starting point for a service management capability.

First, and most problematic, is that these days everyone is an expert. Service, and service perception, can be very personal and almost intuitive in many ways. It is difficult to find people who do not have very strong views as to the kind of service they expect. Unfortunately, although the views may be strong, there is no law that requires that they also be well thought out and clearly articulated. Service provision, being in most respects intangible, can be a difficult area in which to adequately justify what the service requirements are. It is for this reason that much of the emphasis of requirement definition needs to go further than *what* and really evaluate *why* a particular requirement exists. Understanding why a requirement exists must be the result of a real evaluation and understanding of what the users are trying to achieve through the service. This has to be related to the business objectives of the user organization and the business processes being facilitated by the service.

The second problem that has a particular impact on the definition of service requirements in a technology-based service is the very close interaction between system and functional requirements and the requirements of service provision. Both aspects will affect the customer's perception of service, but when defining a strategy for service management it is essential to differentiate between functional and service requirements in order to maintain a clear understanding of where a service organization adds value to the provision of system capabilities.

The third major reason that requirement definition is difficult is that there are a number of different communities, each of which can be considered a community of users of the service. This includes communities as diverse as the service provider staff themselves, user management, and service provider management, in addition to the traditional end user. Each of these communities of interest have different objectives and different angles from which they perceive the service. The diverse nature of these objectives will almost inevitably lead to conflicts of interest regarding the service. Although sometimes tempting, it is not sustainable in the long term to ignore any one interest group. They all have a valid part to play in the service relationship. Part of service management is to ensure that a service relationship between all of the user communities is managed so that the different needs and objectives are understood and accepted across the whole service community.

3.1.1 Questions About Service Requirements

The three problems in defining the service requirements discussed above lead us to three major areas that need to be understood as part of the requirements definition activity:

1. What is the service?
2. Who are the users of the service?
3. What do the users want from the service?

The answers to these three questions are basic to everything that needs to be done to provide a service that will be perceived as high quality. Unfortunately they are not easy questions to answer.

The question "What is the service?" is almost impossible to answer without understanding the answers to the other two questions. Although defining the service may look like a good starting point for answering the other questions, it should probably be the last question to be answered, as a conclusion drawn from the answers to the questions of who the users are and what they want from the service. It is the users and their needs that define the boundaries of the service to be provided. Any theoretical initial definition of what the service is would need to change once the answers to the other questions had been found.

Users of a service come in various shapes and sizes, so every user might represent a unique set of requirements. The key to service provision is to acknowledge that fact and create a generic and common service environment that enables every user to believe they are being treated as an individual user. Fortunately, this doesn't mean that the service environment must be able to handle infinite varieties of user requirements, but it does mean that the user needs to believe that the environment meets their specific needs.

The most obvious means of implementing this concept is through the use of user customization or tailoring options within the system environment. It becomes more complex when moving way from the system environment and implementing service elements on a per-user basis. The need for consistency of service performance naturally leads to proceduralization and standardization of service capabilities across the user community. Certainly in the areas of public service provision there has been a trend toward looking at each corporate user as having a distinct set of individual requirements, and more custom services are being provided than ever before.

3.2 USER WANTS, NEEDS, AND EXPECTATIONS

Requirements for service provision must be understood in terms of whether they are what the user wants, what the user needs, or what the user expects. These three aspects are very different in more than just semantics, and they tend to be determined by the different environmental factors acting on the user. They may at first seem to be beyond the control of even the most comprehensive service management implementation. However, it is important that they are at least understood, and in some cases they may be influenced by the service provider.

What the user wants may not be articulated in terms of a technical capability or service; it is more likely to be related to the business objectives being supported by the service. The service provider must be able to understand these objectives in the terms expressed by the user in order to be able to specify a service that can support them.

The needs of the user are also likely to be expressed in business terms, but probably with a greater emphasis on the capabilities required rather than the objectives to be met. Both wants and needs must be understood and, hopefully, reconciled.

What the user expects is the main area in which influence can be exerted by the service provider, as expectation is a result of definition of needs and specification of how the service can meet those needs.

In order to understand user wants, needs, and expectations, it is necessary to understand the users themselves.

3.2.1 Defining User Communities

Any service has several different communities that have requirements on the service. The first and most obvious is the end user, who has requirements for the functions and performance of the service and the level of help and support available.

Second, the user management has requirements on the service for the management of the user population, the control of costs, and the ability of the service to support the organization in meeting specific business goals.

The service provider staff—the people involved in the day-to-day provision of service—are the third group requiring service. Their requirements are important, because often their needs determine their ability to provide a level of service that will meet the requirements of the end user and user management. One particular requirement that needs to be understood by the service provider organization is the kind of service level that the service provider staff want to be able to provide. If the service provider staff are not comfortable about the service being provided, then it is usually the case that they won't provide it very well.

The fourth community that has requirements on the service is the service provider management and, through them, the rest of the service provider organization. They have requirements to be able to show the effectiveness of the service being provided in meeting the objectives of the organization itself. This community is often forgotten, but in the past I have been closely involved in the development and provision of a service where the ability of the service to meet the needs of service provider management meant the difference between the success or failure of the service due to the level of support it obtained from management within the organization. There have been services that, while achieving a high level of end user acceptance, never adequately met the needs of the service provider management, and as a result eventually failed through a lack of investment and internal support for the service being provided.

None of the above user communities can be ignored by the service provider and in fact should not be ignored by each other. One way in which a successful service can be provided is by ensuring that all user communities involved understand one another sufficiently to realize that the provision and receipt of quality service is a cooperative effort that can be successful only in a partnership between all the user communities involved. This cooperation has to start at the time when the requirements for the service are being defined.

Understanding the communities and the users within them is the main path to effectively defining and understanding the requirements those users will have for any services to be provided. Each community of users has different requirements of the service, and just about the only thing that can be guaranteed is that no two communities have the same set of requirements. Some of them will be complementary to one another, and others will conflict. That conflict is something that the process for defining and understanding the service requirements must be able to manage.

3.2.2 Understanding the Users' Business Needs

Understanding the users starts with understanding what they are trying to achieve; that is, the business function they are using the service for. The gap between user needs and what is actually delivered should be minimal, because otherwise the service does not enable the user to complete their business function. Furthermore, as the business function changes, so may the requirements for the service. This means that a constant liaison between the service provider and the service recipients is required.

Unfortunately, service requirement definition is not a one-time activity, but rather a constant iteration of predicting and reacting to business need. As businesses change, so do the needs of the users from the services provided. The service provider, through a close partnership with the service recipients, needs to be able to predict the new service requirements that a changing business environment will create, and ensure that the service continues to meet business need. Technology represents a potential competitive edge through either cost saving or improved functionality, but it is an edge that can be quickly blunted if outdated service capabilities are applied to meet business needs that have moved on. This means that requirements definition and understanding is an ongoing process. It does not allow a very neat approach to service development because requirements do not remain stable throughout a service lifecycle. Flexibility in service provision is required to ensure that the service adapts to changing business needs.

3.2.3 Understanding the Users' Business Processes

Of course, in addition to the business requirements articulated by the user communities there will also be other requirements that are driven by the individual regardless of the business function. People work in different ways and require services to support their different methods of working.

This particular requirement can be in conflict with the requirements of user management in some industries where technology is of benefit due to the consistency it brings to methods of working and business processes, and is actively being used to re-engineer business processes to support the wider business objectives. In these cases the business objective should take precedence over other requirements. Many of the more individual requirements have no basis in business need and thus represent wants rather than needs. The separation of the "wish list" from the essential requirements that must be satisfied in order for the service to be effective is essential.

However, the wants should not be underestimated or ignored, because often the essential business requirements are interpreted as a given, and it is the failure of a service to fulfill the user's wants that leads to the failure to create user acceptance of the service. It must be acknowledged, though, that not all wants can be satisfied. The gap between wants and the actual service can be managed through managing the expectations of the users in such a way as to ensure user acceptance of the service.

3.3 TRANSLATING BUSINESS NEEDS INTO SERVICE REQUIREMENTS

Any process by which requirements are defined should ensure that the requirements come from the users of the service, across all user communities. The users should not be expected to specify the service required; rather, they should define their requirements in terms of business need. The user and the service provider should work together to translate those business needs into a specification of a service. The translation of requirements defined in business terms to a definition that can be used to implement a service is an area where many problems occur through a lack of communication and understanding between users and implementors. In a technical service definition, as in literature, meaning can be lost in translation.

There is no substitute for talking to the users, even if they don't know or cannot express what they need in terms the service provider can easily understand. The dialogue between the service provider and the user community should try to ensure that a common understanding is reached. Unilateral decisions made by the service provider without reference to the user communities have a good chance of being wrong. Part of the role of the service provider in the provision of service is to ensure that the requirements—and the implications of the requirements—are fully understood by all user communities. This is normally a case of constant iteration and refinement, but the process can be minimized if each party can provide an understanding of why a requirement exists rather than just what the requirement is. To do this the service provider must have an understanding of the business of the users, and the users must understand the issues of concern to the service provider. Again, the understanding of *why* rather than *what* greatly increases the chance of implementing a successful service.

3.3.1 Managing Conflicting Requirements

The requirements process must enable a clear definition of requirements, with a business justification that can be understood by all user communities. This will enable the rational prioritization of those requirements against one another. The prioritization should be agreed upon across all user groups and defined with adequate consultation and understanding of why requirements have been prioritized in a certain way. There may be instances in which requirements conflict, either implicitly through resource constraints or explicitly through the needs of different user communities being directly opposed to one another. This is not necessarily a problem with the service, but is instead a more fundamental problem with the relationship between the user communities. Conflicting requirement are just a symptom of the problem. When this occurs, the basic business conflict should be addressed, rather than treating the conflict as just a technical or service definition problem.

Perhaps the most common conflict is due to lack of resources to implement the service capabilities defined by the user communities. This conflict must be managed through the management of expectations and the management of requirement prioritization. As long as those two elements can be controlled, it is always better to meet some of the requirements well rather than all of them badly. Meeting a requirement badly is usually worse than not meeting the requirement at all, because if the user has an expectation that the service will not meet a specific requirement, then they will normally find another way to do something. If the expectation that the service will meet a requirement is not fulfilled, then this will negatively affect the user's ability to fulfill their business needs. The understanding of requirements is key to enabling the service provider to do what the user community expects the service provider to do. Meeting customer expectations is the key to creating customer satisfaction.

4 Measuring Customer Satisfaction

Services exist only in order to fulfill the functions required by the users. A service that fulfills the desires and expectations of the users will be successful; a service that does not meet user expectations will fail. The implementation and management of a system and process to understand and then manage customer satisfaction levels is an exercise that should be at the heart of all service management organizations.

There is a view that the only true measure of how effectively a service is being provided is the satisfaction level of the customer or user of the service. There is an alternative view that suggests that customer satisfaction is a fundamentally flawed method of measuring service performance because it is subject to the vagaries of personal opinion as well as so many other indeterminate factors that it can never reflect an accurate view of service performance. As with most extreme and polarized views, neither is completely right.

To be sure, customer satisfaction is very difficult to measure and monitor, but there is great value to understanding the level of customer satisfaction being achieved. Measuring customer satisfaction is a check on the future success of the service. Certainly in the provision of public services, the only reason customers will part with their money for the use of a service is if the service adds value to their business.

There are many things that can be measured as an alternative to customer satisfaction, such as service availability, response times, number of faults, length of time the telephone rings at the help desk, and so on. I wouldn't want anyone to stop monitoring and striving to improve any of the above measurements of service capability, but none of them represents an end in themselves. While all very worthwhile, they are measurements of items that determine customer perception of service quality and not direct measurements of the customer's perception of the service or the customer's satisfaction. The concentration on such measurements is a little like ensuring that all the players on a football team are playing well, without bothering to check whether the game was won or lost. Customer satisfaction is more than the sum of the parts. While good performance by each individual element may be indicative of successful service provision it is not a measure of the real test of customer satisfaction.

For example, a service could be operating at 99.99% availability, but if that 0.01% unavailability was when a particular user wanted to access the service, it is unlikely that the user is 99.99% satisfied. They are probably extremely dissatisfied with service performance—all they saw was 100% unavailability.

It is entirely possible for all indicators to suggest that service performance is acceptable when there is actually a latent customer satisfaction problem that will affect the service in the future. Sadly, it is not possible to rely only upon customer complaints to indicate when satisfaction is low (despite indicators suggesting otherwise), because not all dissatisfied customers will complain. There are many reasons for this, ranging from not knowing who to complain to, to simply not using the service rather than bothering to complain. A silent customer is not necessarily a happy one.

This is the reason why customer satisfaction is such an important measure of service success in the present as well as in the potential for the future. There is no other way for the service provider to understand the level of service acceptance and the reasons why a service is accepted or not accepted by the user population. There is nothing quite as clever as a user trying to find an alternative way of doing something when the service lets them down. Their ingenuity knows no bounds, and the service provider never gets to know about it until it is too late to react and correct the problem.

However, customer satisfaction is very difficult to measure and even more difficult to manage, since, as we said earlier, it is the sum of a large number of disparate elements, not all of which are under the control of the service provider.

4.1 TALKING TO THE CUSTOMER

There is only one way to measure customer satisfaction, and that is to ask the customer. There are other measures that are an indication of satisfaction such as trends in the usage of the service, but these trends often do not become meaningful until it is too late. The trend can be hidden by increases in the numbers of users, and by a certain amount of lock-in of existing users into using the service despite their dissatisfaction, while they devise an alternative way of achieving the same objective. By the time these trends become obvious, the damage has been done. Users who stop using a service are unlikely to return to it once they have invented their own way of doing things.

Proactive measurement of customer satisfaction through direct contact with the customer and the user enables the service provider to identify trends in service satisfaction levels before the users have identified alternatives to the service. It also has a useful byproduct: giving the user the ability to express dissatisfaction will discourage them from identifying alternatives. Once the user's opinion has been made known, and as long as there is some faith in the service provider's willingness to resolve the problems, an improved existing service is an alternative itself in the mind of the user.

4.2 CUSTOMER EXPECTATIONS VERSUS CUSTOMER EXPERIENCE

Customer satisfaction is determined both by the expectation of the customer and by the experience of the customer with regard to that expectation. This means that the process for managing customer satisfaction starts well before the customer ever becomes a user of the service involved. This applies to all user communities, including the staff and management of the service provider, as well as the end users of the service. For example, say the service provider management have been given an expectation that the cost of providing a service must be a certain amount, but the actual costs overrun. The level of satisfaction that the provider management have is likely to be less than if the actual costs are the same as the predicted and budgeted costs of providing the service. If the actual costs are less than those budgeted, then satisfaction could soar to previously unknown levels.

4.2.1 Setting Realistic Customer Expectations

To set expectations correctly it is necessary to understand the reasons why the user communities are involved with the service, that is, their business objectives. These business objectives will determine their expectations of the service. Thus the process of gathering, defining, and evaluating requirements is also an integral part of the process of managing satisfaction. The requirements process defines the objectives and therefore the expectations that the user communities have of the service. It is the starting point for user satisfaction.

From this starting point of customer expectation, the basic parameters of customer satisfaction have been established. The expectation of the user is the baseline for what will determine a satisfied customer. Certainly in public services this is a major cause of future service satisfaction problems. In order to make a sale it is often the case that customer expectations are set at an unrealistically high level. Once the salesperson has banked their commission, the rest of the service provider organization is left to manage the customer's expectations back to a realistic level. This can be a very painful process for all concerned (including the salesperson).

For public service provider organizations the differential between the sales message and the service delivered is a complex problem, the causes of which normally predate the salesperson's incorrect setting of expectations. The chances are that in many cases the salesperson's expectations have also been incorrectly set by a product marketing function trying to motivate a sales force with several products or services to sell. Even prior to that, the problem is almost certainly rooted in a lack of understanding of business requirements and the limiting factors of technological capabilities. Unrealistic expectations probably exist in the minds of the customer before a salesperson has ever spoken to them. It is always much easier to say yes early in the sales cycle to ensure a sale is made, then wait until a decision to change providers is costly and difficult for the customer before attempting to reset expectations to a realistic level.

The same sales process problems can also take place within the service provider organization, when investment funds are limited and several product development functions are competing for these resources. It is not unknown for unrealistic rates of return to be shown within the investment case. Margins that are normally only dreamed of can be justified, and a net profit exceeding the gross national product of the combined markets the service addresses magically appears on the bottom line. Unrealistic expectations are the curse of all services and all user communities.

The correct expectation has to be set at the beginning of any service implementation, through the definition and clear understanding of the requirements that service is trying to meet and how it is going to meet them. As the service is developed, there is a need for constant, almost tedious, iteration to check and recheck that what is being implemented is meeting the requirements that are defined. This iteration must be done with the user community, rather than on their behalf by the people who are developing the service. Only by the constant and ongoing involvement of user communities in the definition of their service can the reality of the development not diverge from expectations. This is a responsibility on both sides of the service partnership, not just a service provider responsibility. Any user community not willing to invest time and resources in verifying that their expectations are being met will almost undoubtedly get the service they deserve.

Setting expectations is only the first stage of managing customer satisfaction. Service provision is a new challenge every day the service is provided. The potential to create a satisfied or dissatisfied customer exists every time a user has contact with the service. The management of customer satisfaction is an issue for every activity, every day.

4.2.2 Managing Customer Satisfaction

The ongoing management of satisfaction is an exercise in ensuring that the actual service delivered and the expectations of the service delivered continue to be synchronized. Again, the only way to find out whether this is the case is to ask the users of the service, and only to ask them not once, but to maintain a continual program of verifying customer satisfaction on a regular basis. The frequency of verifying customer satisfaction is required because service provision is an ongoing activity. Both expectations and experience vary over time. In the case of changing experience this can often be determined purely by the latest experience of the service.

Memories of service received are relatively short. In particular, previous instances of good service are easily displaced in the user's mind by one instance of service that fails to meet their expectation. The need for consistently good service provision is very strong. If the quality of the service delivered is erratic, then both user experience and expectation can fluctuate wildly. This makes the balance between experience and expectation more difficult to manage.

Expectations are also built up over time. It is ironic that customer satisfaction problems can be *caused* by the provision of high-quality service—service above committed or initially expected levels of service. This will create an expectation in the user community that is unsustainable.

A service with which I was closely involved was based around a computer-based interactive application. This application had a committed response time of two seconds for any user interaction with the service. This was understood by the user communities, and the service operated satisfactorily for some time. As usage increased it was decided to upgrade the mainframe processor the service ran on. The upgrade implementation went well, and with a more powerful processor response times fell to about 800 milliseconds, greatly exceeding any expectation set with the users. No complaints were received at this time. As usage volumes again increased the response time grew closer to the committed two seconds with which the users had previously been perfectly happy. Unfortunately, having now experienced 800 milliseconds, the users were no longer satisfied with the two second response time, and customer satisfaction with service performance took a considerable downturn, despite an improvement in performance and considerable investment in improving the service. What we had done was reset user expectation to 800 milliseconds—an expectation that could not be met in a sustainable way. This was a lesson learned the hard way.

As expectations and experience change, the level of satisfaction with a service also changes, and needs to be continually verified with the users. This process of verifying customer satisfaction with the users is obviously subject to all the concerns regarding the measurement of customer perception of a service: it is inexact and prey to an immense number of affecting factors, some of which are outside the control of the service provider. However, the value of measuring customer satisfaction exceeds all the associated problems. The process used to measure satisfaction must be defined in such a way as to minimize the effect of the problems with measuring satisfaction.

4.3 METHODS OF MEASURING CUSTOMER SATISFACTION

The main way in which the user community satisfaction levels can be judged is through the use of a customer satisfaction survey. The construction of the survey questionnaire and the way in which the survey is then conducted can have a significant bearing on the results that are presented and the value of those results to the process of managing customer satisfaction.

Both the scale and the format of the survey undertaken will vary according to the type of service that is being offered and the type of user being surveyed. However, the survey should be seen from the outset as more than just the collection of data. The process must include the analysis of that data, the action taken as a result of the analysis of the data, and the verification and communication of both the analysis and the actions resulting from the data collection exercise. Any satisfaction survey process is not and—just as importantly—must not be perceived as just an exercise in data collection. Users must believe that their views will be listened to and will cause actions to be taken by the service provider where necessary. This makes the exercise a valuable process for both parties.

Asking the right questions is a key success factor in any customer satisfaction survey. There is not much point in discovering how satisfied the user is with elements of the service that are not important to that user. Any meaningless exercise is only likely to lead to dissatisfaction with the satisfaction survey—not the desired result at all. The relevance of the questions being asked should be made clear, and the survey must be built on a clear understanding of the elements of the service considered important by the users. The involvement of the user community in the creation of the satisfaction survey would be an ideal way of achieving this.

The collection of the data for the satisfaction survey can be accomplished in several fairly traditional ways, such as by written questionnaire, telephone questionnaire, or face-to-face meetings. The method is normally dependent on the number and type of users being questioned. Clearly, face-to-face meetings with several thousand individual customers is not feasible. For services with many customers, the written questionnaire is probably the easiest and most convenient method of collecting information, as it can be completed at the user's convenience, rather than as an interruptive telephone call.

The format of the information gathered should enable simple analysis to be done quickly. Too many satisfaction surveys collect data, but any follow-up action is either not done at all or done so long after the data collection exercise that it is no longer of any relevance. The rating of aspects of the service on a scale of one (excellent) to five (unacceptable) is a simple way of gathering easily analyzed data.

Ensuring that the data collected is useful for determining future actions to maintain or improve satisfaction requires that the sample of users taken is correct. It is not adequate to approach a random selection of users. The opinions of one group of users should not be allowed to dominate the results of the survey. It is also not advisable to use the same set of users for the opinion survey feedback each time a survey is conducted. Changing the group that is asked to complete the survey prevents the survey itself from becoming a burden on the users and also ensures that a wider view of the service is obtained in terms of both the user base and the time at which the survey is taken. The fact that a user may be surveyed, then perceive improvement, and then be surveyed again may affect the feedback received in the second survey. It may be more accurate to implement improvements on the basis of survey feedback from one set of users, and then ask a fresh group for their perceptions of the service.

Once the data is collected and put into an understandable format, the results should be made available to all who participated in the survey. Honesty and openness at this time is important. If the results of the survey are bad, there is little point in hiding this from anyone who participated, as they provided the information that led to a poor result. They obviously already know that the service is poor.

The next focus must be on the creation and promulgation of an action plan to resolve any issues discovered through the survey. Again, the creation of an action plan in conjunction with those who determined that action was required will ensure user involvement in the process and provide credibility in the eyes of the users for the actions being taken.

Once these actions are complete, the job starts all over again.

4.4 CONCLUSION

The achievement of customer satisfaction should be a central objective of all service organizations. Although complex to define and manage, it is almost certainly the single most valid measure of the whole service provider organization involved in setting and meeting expectations and creating customer perceptions. Ensuring focus within the service provider organization will require that satisfaction is measured and monitored and that the satisfaction of customers is an achievement measurement for all involved in the process, through bonus schemes, annual appraisals, and other performance measurement means. The continued satisfaction of customers is what will make any service successful in the long term.

5 The Service Partnership

The important element of the satisfaction survey process is that it is undertaken as a form of partnership. Both sides of the service relationship participate in the exercise and both derive value from it. As with the satisfaction survey, almost all elements of service management are better implemented as a form of partnership.

The provision of service is based around the relationship between the two sides of a service partnership, the service provider and the service recipient. In the past this has almost invariably been treated as a normal customer and supplier relationship. However, a service relationship requires something better than this strict demarcation of roles and responsibilities and an almost adversarial positioning. It requires a true partnership to be formed between the service provider and the service recipient.

A service relationship is one that is ongoing. This is very different from a normal customer and supplier relationship, where the relationship is a temporary one lasting just long enough to transact a sale. It is the ongoing nature of this relationship that requires a different approach. Each party to the relationship has rights and responsibilities to ensure that relationship works on a day-to-day basis. Neither the rights nor the responsibilities belong to only one side of the relationship.

In the provision of service it is important to acknowledge the nature of this ongoing relationship, as it determines the precepts behind the relationship and has a material effect on the way in which the two parties work together. If two organizations view one another strictly in customer and supplier roles, their approach to the relationship will be very different to two organizations that determine that they are in partnership with one another for the provision and receipt of services.

In a partnership of this kind, more than just the provision of goods for reward is required. Both sides of the relationship are involved in ensuring that the relationship works effectively for both parties. The customer-supplier relationship can be adversarial in that the customer requires more for less from the provider. Although this is a fact of life to some extent, the partnership determines that this adversarial approach is minimized, because each party has a vested interest in the success of the other.

In some service relationships with which I have been involved, the service recipient has dominated the service provider to ensure the provision of services to the customer in a certain way and at a certain cost. In the short term this resulted in certain benefits for the service recipient while being detrimental to the service provider. In the long term it was detrimental to both parties. The cost base of the service provider was increased by this customer's requests, and the efficiency of the provider organization was damaged by the need to constantly react to a difficult customer. In the ongoing provision of service it is in the best interest of the service recipient to create an efficient service provider. In the same way, it is in the best interest of the service provider to create an efficient service recipient. Every day, the two parties affect one another, and if one side of the relationship is ineffective it ensures that the other party is made equally so.

The customer-supplier model does not work well in the service environment because of the impact each side of the relationship has on the other. In some ways the relationship changes in service provision, because at any one time either side could be the service provider, while the other is the service recipient. A prime example of this is in the area of fault management, where a problem has emerged with the service and the recipient requires the problem to be resolved by the provider. A simple view would be to see that the responsibility for the fault resolution lies solely with the service provider. This may not be accurate. Initially, in the fault resolution process, it is likely that the service provider is dependent on the service recipient to report the fault and to be able to describe the symptoms being experienced. The quality and timeliness of the information provided by the service recipient will directly affect the ability of the service provider to resolve the fault. Both sides have to act in partnership, providing service to each other, in order to facilitate the resolution of the fault. Service is a two-way street.

Acknowledging the partnership nature of the service relationship makes it possible to better define the requirements on both parties to make that relationship work. Too often service is perceived as operating in only one direction—from the service provider to the service recipient. This is where the traditional customer and supplier relationships are applied to a service environment. This approach is damaging in a service environment because it does not encompass the important aspects of a service relationship. It does not allow for the mutual interdependencies for working together on everything from the requirements definition to fault management. This partnership is what can make a service relationship work for both parties.

Future service developments from service provider organizations need to be designed to further break down any firm divisions in the customer-supplier relationship. In network services, advances in this area vary from the relatively simple approach of making network management capabilities (normally the sole preserve of the network provider) available for use by the service recipient to enable them to manage their use of a public network as if it was a private capability (virtual private networking), to more complex organizational initiatives such as "virtual organizations," in which customer and supplier staff work together in a defined virtual organization incorporating staff from both real organizations.

Both of these examples of partnership initiatives are valid implementations of the same basic underlying partnership philosophy: service delivery is a partnership. There will always be differences in the level of partnership that is appropriate for any single relationship between a service recipient and a service provider. These differences are determined by a variety of factors based on the attributes of the organizations involved. No single model can be equally applicable across all organizational relationships.

5.1 BASIC REQUIREMENTS OF A SERVICE PARTNERSHIP

Before being able to determine the most appropriate partnership model for any two organizations, there are a series of basic requirements that must be met by both organizations as the baseline for entering any partnership relationship. These requirements apply equally to both the service provider and the service recipient side of the relationship.

In many ways these basic requirements for a partnership service relationship acknowledge that the determining factors in a successful partnership are very much based on the personal relationships that are built across two organizations. As a result they are very similar to the personal attributes that determine the success of any relationship based on partnership. In addition, whenever a strong organizational culture or identity exists that is distinguishable from the personal relationships involved, it is important that the same attributes are determined to exist at an organizational or cultural level as well as at a personal level. The organization must have a set of values it promotes and adheres to in the way it operates.

5.1.1 Mutual Acknowledgment of the Partnership

The main requirement that exists is for both parties to acknowledge that the relationship is a partnership and as a result is symbiotic in nature. The success of one party very much depends and is determined by the success of the other participant in the partnership. This means that both sides must work together on the basis that no long-term benefits can be gained from the relationship at the expense of the partner. The only benefits that can be achieved selfishly are those that will damage the relationship in the long term despite an appearance of a short-term gain. When the relationship is damaged both parties are the losers.

For example, as a customer, an objective may be to get the best possible price from a supplier, regardless of the benefit of that price to the supplier. However, a bankrupt supplier is of no use whatsoever to a customer in the long term. The short-term benefit of a lower price in partnership between service recipient and service provider should be outweighed by the long-term benefit to the service recipient of having a viable partner and a viable relationship.

The example above is somewhat black and white in order to clearly identify the problem. In the real world the problem is much more subtle and is an extremely hard one to come to terms with. Customers who are perceived as being difficult or whose actions are not supportive of the goals of the service provider are unlikely to receive quality service from the service provider. There needs to be a commonality of goal and mutual support for one another's objectives for the relationship to be successful. I know from experience that if support staff have a list of problems to deal with, the last one to be handled will be the one in which the customer is unpleasant to work with. It's not that surprising when you think about it—after all, nobody goes to work in order to be abused and mistreated by people they are trying to help. Difficult customers do not get the best service; his is determined by nothing more than human nature. Above all else, service is a very human business.

The relationship between provider and recipient becomes even more complex when a shared service is being provided by a single-service provider to multiple service recipients. In this environment the management of the relationship is very complex because, through the service provider, an indirect relationship has been established between multiple service recipients who may have nothing else in common except for the fact that they are receiving service from the same provider. However accidental this relationship may be, it is important for all parties to understand that the relationship exists and that it is essential in a shared service environment to manage the indirect relationships as well as the direct relationship.

In this environment it is an unfortunate fact that multiple service recipients can in some circumstances view the relationship as a competition for the resources and attention of the service provider. Since the relationship between service recipients is indirect, the competitive nature of the relationship shows itself in the relationship between the service recipients and the service provider. Service recipients rarely exert pressure on each other. Pressure from a service recipient to receive a level of service superior to the service provided to other recipients is one manifestation of the competition. However, as soon as a service recipient attempts to do this they are immediately undermining the partnership relationship that the service provider has with other service recipients. This then damages the service provider's ability to meet its own goals by maintaining all the relationships that are important and, in the long term, undermining the specific relationship that the service recipient is trying to maintain at a superior level. By damaging the service provider's relationship with other service recipients the service recipient ultimately damages their own relationship with the service provider as well.

The dilemma facing the service recipient is one of trust. If all service recipients and the service provider are operating under the same rules and with the same objectives for all participants in the relationship, then there is nothing to worry about. If just one party to the relationship does not abide by the rules then the relationship is ruined for all, with the party who damaged the relationship probably achieving some short-term gain above all other participants. In the long run everybody loses out.

5.1.2 Mutual Trust

Trust is one of the basic attributes of a partnership relationship between service provider and service recipients. This trust operates on several levels: both personal trust and trust in the organization. Furthermore, it is not just a discussion about the service recipient trusting the service provider—the service provider also needs to be able to trust the service recipient. This is often forgotten, especially by the service recipients.

The required trust covers several areas, most of which are applicable to all service relationships, but each relationship will have its own specific aspects to it that will also need to be trusted by the parties to the relationship. Both sides must trust in the commitment of each other to the partnership. A service provider almost invariably needs to make an initial investment of resources and time in providing a service to a specific recipient. The service provider normally aims to recover this initial investment over the lifetime of the relationship. If the relationship is a temporary one that does not enable the service provider to recover that initial investment, this is not supportive of the business goals of the service provider. Both the service provider and the service recipient need to regard any service relationship as being for a significant period of time.

On the other side of the coin, when service recipients choose to move to a service provided by another organization, there is always an element of lock-in of that service recipient to the service that was previously being provided. There is also an initial investment on the part of the service recipient in terms of the time and effort and usually business disruption that goes hand in hand with implementing a service within an organization. Any new service requires an investment by the service recipient. In the public service environment there are normally initial implementation charges to be considered. In any service environment there is the investment associated with the service definition and development as well as user education, training, and documentation. It is still not as easy to change service providers as it could be. It is possible that this is a symptom of the fact that few genuine service partnerships have been established. In order to ensure that service provision is profitable, a degree of lock-in may be desirable for the service providers.

Once the user community is dependent on that service and has been educated to use that particular service, it is a costly and at times painful exercise to try and move to a different service provider. The service recipient needs to be able to trust the provider to be committed to the delivery of quality service on an ongoing basis. It is also in the interest of the service recipient to foster the commitment of the service provider.

As in all relationships two of the main elements of building trust are honesty and openness. It is essential in the service environment that the provider and the recipient are honest with one another. If service quality is poor it is important that both the recipient and the provider acknowledge this. By the same token, both must acknowledge when service quality is good. In order to effectively target resources at the right elements of the service, the service provider must be allowed to understand what is important to the service recipient. If, for example, a problem is manifested within the service, the recipient must be honest about the impact of that problem on his business goals and objectives. It benefits no one if a low-impact problem is consuming resources that could be of greater benefit elsewhere.

5.1.3 Rights and Responsibilities

The balanced nature of an effective service partnership can be embodied within an understanding that both the provider and recipient have certain rights that they can expect from the relationship. At the same time they have certain responsibilities that are associated with those rights. There are no free lunches in a service partnership, and every right normally carries with it a responsibility.

Although these rights and responsibilities will differ depending upon the service being provided, it is an essential part of the partnership that these rights and responsibilities are understood. Contractual agreements and service level agreements are both being used by service providers to define rights and responsibilities in areas such as fault reporting, system performance, and change management. Again, these rights and responsibilities need to be understood by all areas of the organizations involved, so that the individuals exercising the rights or fulfilling the responsibilities understand exactly what is expected of them within the service relationship.

Contracts and service level agreements are only half of the story. Both these elements often require lengthy negotiation and invariably the involvement of the legal departments of both the service provider and the service recipient. This doesn't always make them the most effective mechanism for managing the day-to-day issues of service provision. They do, however, provide an essential framework for the further detailed definition of rights and responsibilities. Ideally the relatively high-level definitions contained within a service level agreement, for example, would be expanded by the provider and recipient to drive the development of joint operational procedures that would ensure that rights and responsibilities are documented within the definitions of day-to-day activities.

While the high-level agreements do provide the framework, the relative rights and responsibilities need to be understood by the people who deal with each other on a regular basis. If everyone understands what they are responsible for in the relationship, their interaction has a common understanding to work from. This is particularly important in areas such as change management and fault management. Changes to a service environment can create severe problems if people are not aware that a change is taking place or are not aware of the impacts of the change and have not prepared and planned accordingly. The agreements and understanding must filter through the whole organization of both the service provider and the service recipient.

The roles and responsibilities associated with the agreements are normally embedded within the organization and the management systems that support the organization. The structure and systems of both organizations, and how they relate to each other, have a major impact on their ability to deliver day-to-day service.

The concept of a "virtual organization" encompassing both service provider and service recipient staff within a virtual organization structure is one way of ensuring organizational interlocking for the delivery of day-to-day service. This is a particularly innovative approach and may not be possible in all circumstances, but achieving a focus on the delivery of service through the organization and management systems is an essential part of a service management implementation.

6 Organization and Management Systems

The organization and management systems of both the service provider and the service recipient organizations have the potential to either encourage the delivery of quality service or act as a very effective blockage to all the best of intentions. While good organization and management systems can never make up for individuals who deliver poor service, bad systems can hamper good people to such an extent that service perception is poor whatever they do.

As with many aspects of service management, there is no single right answer to the question of how the service organization should be structured or what systems should be in place. There are some basic building blocks, however, that can help to ensure that an organization develops in the right direction.

6.1 FOCUSING THE WHOLE ORGANIZATION ON SERVICE PROVISION

The first and most fundamental organization system starts with a service focus developed for the organization as a whole. The service focus can be established by setting the direction of the whole organization to achieve the goals of customer satisfaction and service quality.

The focusing of a whole organization on service provision can be very difficult to achieve as this requires that the day-to-day goals and objectives of every employee are focused on service quality. While this may represent an ideal, there are several conflicting pressures on many aspects of an organization. Cost and available time and resources are major pressures in almost all organizations, and these are frequently a trade-off against service delivery. Although quality may be free over the lifetime of a service, at certain stages of service development the costs of quality are very clear in terms of time and resources.

Within any service organization, there are very disparate functional areas ranging from service support staff to finance and human resources. Trying to organize all these elements, many of which are traditionally several steps removed from any customer facing role, is a task of heroic proportions. The organization that achieves it will have a distinct advantage over those that give up. In order to focus an entire information technology organization on quality and service provision, there are some approaches that can be applied across the organization to make everyone concerned with service provision.

6.1.1 Incentives

The incentives attached to all employees within the organization can be related to the quality of the service and the customer's perception of the service. This can be achieved to some extent through bonus payments related to the measured satisfaction of customers. However, day-to-day measurement of satisfaction is difficult to achieve, and basing bonuses on an annual customer satisfaction rating can tend to make the relation of satisfaction to pay a remote aspect of the organizations focus every day. Incentives really need to be at the forefront on a daily basis, rather than just as an annual event.

Daily measurement of the elements of service that set customer satisfaction levels—such as problem resolution times, service availability, and service performance—must be extremely sophisticated in order to ensure that the measurements can be accurately translated into a measurement of performance that could be used for incentive purposes.

6.1.2 Roles and Responsibilities

The roles and responsibilities of each individual involved in the delivery of service need to be clearly defined in order to ensure that the focus on the customer is contained within that role. Roles and responsibilities should be formalized within each employee's job description.

Definition of roles and responsibilities is a key element of ensuring service delivery. While it is difficult to define service delivery elements within areas normally removed from the customer, such as finance and other support functions, it is possible to define roles with a focus on the end result for the organization. For instance, one of the recruitment responsibilities of human resources could be to ensure that service-oriented staff are put in place. It's important that this responsibility is placed within a service organization, as there is always the potential for the "tail to wag the dog" and for support functions to hinder rather than help service delivery.

6.1.3 Job Objectives

The job objectives on which employee appraisals are based also should contain quality and customer satisfaction elements. It is often the case that a series of job objectives that address the function of the role are then enhanced by an additional objective to address service delivery. It is preferable that service delivery elements are not identified separately in a set of job objectives but are explicit within all other objectives.

I am also very much in favor of including wider objectives within individual job objectives. This ensures that an individual is measured not only on specific activities, but also on his or her contribution to a wider set of team and organizational objectives. This helps to focus the mind on more than parochial interests.

6.1.4 Mission Statements

The quality and customer satisfaction focus can also be shown through departmental mission statements that define both the role a department performs and the manner in which it will be performed.

Missions and objectives represent a hierarchy of tools that can be used to maintain the focus of the organization. The mission statement of the department would drive the definition of roles and responsibilities within the department, which in turn would generate the job objectives for individual employees on which appraisals would be based and bonus and incentive payments made. The important key to this approach is that the focus on service delivery permeates the whole organization and every individual.

6.2 ASSIGNING SERVICE PROVISION TO CERTAIN INDIVIDUALS

An alternative to any attempt to focus the whole organization on service delivery is instead to identify specific people within the organization with responsibility for these aspects of service. This approach has the advantage of enabling a small group of people to be very heavily focused on service delivery, but it can mean that other areas of the organization might view it as not being their responsibility at all. Ownership of service quality and customer satisfaction should, in my view, be clearly felt by all members of an organization.

6.3 COMBINING THE TWO APPROACHES EFFECTIVELY

A hybrid approach to the implementation of a customer-satisfaction-oriented organization can be the best approach in some circumstances. This requires both the implementation of service quality focus across the organization and the identification of key service delivery roles within the organization. In order for this method to operate effectively, there is an even greater need for the clear definition of roles and responsibilities within the organization.

6.3.1 Responsibility for the Provision of Quality Service

Responsibility for the provision of quality service means that any failure to provide a certain quality of service is directly attributable to an individual. In the hybrid organization, responsibility for the provision of quality service must continue to lie with each and every member of that organization and should be recognized through mission statements and the job objectives developed within that organization.

6.3.2 Accountability for Quality Service

Accountability for quality service requires an individual to be responsible for measuring the quality of service and for identifying when, where, and why it has failed. Specific individuals can be identified as being accountable for quality service on the condition that they alone will not be held responsible for the quality of service. As long as the underlying responsibility is felt by the whole organization, the use of specifically accountable individuals can increase the organizational focus.

6.4 KEY PROCESSES

In order to manage the effective delivery of service, there are some key processes that need to be clearly understood across the organization. When responsibilities cross, and when accountability and responsibility are incorporated in different areas of the organization, it is particularly important that these processes work effectively.

6.4.1 Review and Approval Process

The review and approval process is vital when specific service delivery actions are to be taken or when changes are made to the way in which service is delivered. The review and approval of those changes must be performed in such a way as to ensure that all areas of the organization understand the responsibilities associated with the changes and what they must do in order to meet those responsibilities.

This focus has to be maintained at two levels. First, approval of a change means only that the responsibilities are understood and accepted. Second, disapproval may relate only to one's specific areas of responsibility. For example, an individual responsible for the implementation of new users onto a service should approve changes on the basis that they can continue to implement new users. The individual would not be allowed to withhold approval on the basis that they do not believe that another area of the organization will be able to meet their responsibilities.

This may sound like common sense, but an unclear review and approval process an cause a lot of problems. It's particularly important in a service environment for two main reasons. First, you can't afford to have anyone being half-hearted about any aspect of managing the service. If people aren't fully committed to actions on the service it is almost guaranteed that something will go wrong. The second reason is that change is becoming more frequent in the service environment and, as a result, services need to be more flexible. At the same time, the demand is for even more reliable services. In my experience, most reliability problems are caused by change to the service. If service changes are to be managed correctly, without impact to service quality, then an effective review and approval process is more essential than ever.

6.4.2 Issue Resolution

In any matrix organization there are times when conflict will arise. If, for example, an individual has responsibility for a specific customer who was dissatisfied with the operation of the service, they may request a change to that service to improve the satisfaction of that individual customer. At the same time someone responsible for the delivery of that service regardless of which customer is involved may not agree with the proposed change. It is important that the disagreement can be resolved quickly and effectively to meet the needs of the customer.

To achieve this the management system must make clear the escalation process by which issues are resolved. The escalation process should have clear lines of escalation up a management chain to an understood point at which authority across both sides of the disagreement is available. This element is key to the management structure in that the meeting of an escalation between two areas of the organization with conflicting requirements and needs should not be so close to the initial issue point as to make the escalation meaningless. Escalations within a single first-line management department should be avoided wherever possible. The goals of that department should be such that there is a commonality of purpose that would limit the number of areas in which conflict could occur. At the same time, two related areas of the organization should not be so far apart in the management structure that every escalation results in the managing director's being asked to resolve a dispute.

Service management is often about compromise and resolving conflicts. In many cases decisions are difficult because there is no one right answer. The two keys to resolving an escalation are to achieve a quick resolution of the problem and to ensure that everyone involved will support the decision and implement the solution. Keeping everyone on the same side is an art that every service manager has to learn.

The effective resolution of issues through the management system has a direct bearing on the service delivered to the customer. Some service providers view the organization and management systems as being internal to the organization and not of significant importance to the customer. However, every time the customer interacts with the service provider, the organization and management system affects the level and quality of service received.

For example, if the roles and responsibilities within an organization are not clear, there is a real danger that customer problems will not be effectively resolved. Too many times, customers are passed around an organization as each person with whom the customer has contact denies responsibility for a particular aspect of the service. If the organization is unable to quickly and effectively resolve internal conflicts, the service provided to the customer suffers as the organization tries to resolve its own problems at the expense of the customer.

The management system must enable a customer need to reach the right place within the organization within the right time frame. Then the organization must be able to react to that customer need quickly and effectively.

6.5 SERVICE DELIVERY STRUCTURE

The pressures of cost and control on an organization must be taken into account when defining the organizational structure that is needed to respond to customer requirements for day-to-day service delivery. The organization structure should be based around the needs of the customer, ensuring that each of these needs are identified and that responsibility, authority, and accountability for each of them are clearly embodied within the organization.

The two customer requirements from a support structure are (1) to be able to obtain a clear and effective resolution to any change, problem, or information request and (2) to ensure that the service is maintained and developed in such a way as to continue to meet their business needs as they change. Every element of the organization has a responsibility for determining customer satisfaction and maintaining service quality. However, the different aspects of service management need very different skills and attributes from the staff involved. In order for service management to meet these customer needs, a structure can be developed that provides the following:

1. A single interface point for all queries relating to the service relationship on a daily basis. This interface will be responsible for taking customer calls and ensuring they are managed to the satisfaction of the customer.

2. A structure for the resolution of complex problems that cannot be resolved through a single interaction with the customer.
3. An organizational element designed to ensure that ongoing customer satisfaction is maintained and that the service develops in line with changing customer needs.

It is essential that these two needs are managed separately. Giving strategic responsibility to the same group of people who have day-to-day responsibility is a good way of ensuring that strategic needs are never considered, because today's problems always seem more important.

6.5.1 The Service Manager

Some organizations have within the service delivery organization a service manager who has specific responsibility for understanding the customer's strategic relationship with the service provider in terms of the ongoing provision of service. The service management of a customer should be a separate function from the sales or account management of the customer. Sales and account management will tend to be revenue rather than service driven—focusing on new opportunities rather than the delivery and improvement of current service. It is often impossible to reconcile the two in the minds of the customer even though future opportunities are often driven by the provision of current service. Repeat business is very important to all public service providers. The service manager is one way to generate this.

In some ways it is possible to view the service manager as the customer's representative within the service delivery organization, or a customer champion. The number of different customers with different requirements and demands on the service provider organization means that a voice representing the customer needs to be located squarely in the heart of the service provider organization. It is in the interests of both the service provider and the service recipient to ensure that this level of representation exists.

6.6 CONCLUSION

The willingness of a customer to invest in the organization providing service shows how important this element of service provision can be. Many of the most irritating problems in dealing with a service provider can be the inability of the customer to find the right person to talk to or to reach someone who can make a decision. At times this can be because organization is poor and decision making is not delegated sufficiently close to the areas of the service provider in direct contact with the customer, but it can also just be a lack of communication and understanding. Ensuring that both the service provider and the service recipient understand each other's organization and that these organizations are able to support one another in the service partnership is one of the critical success factors for the provision of quality service. This understanding has be to developed through the whole service relationship. It means that expectations and perceptions have to be managed and that both organizations have to understand each other's values and principles, as well as their organizational structure and management systems.

There are extreme examples in which customers and service providers have worked together in the provision of services as a "virtual organization." The provision of service is achieved through the creation of an organization consisting of both service provider and service recipient staff. Although there are significant cultural problems in the integration of two differing organizations, they can be overcome if the aims and objectives of the virtual organization are understood and supported by both sides.

7 Service Level Agreements

The service level agreement (SLA) represents one of the most powerful tools in the process of setting expectations and perceptions with the customer. However, this potential is often wasted and, as a result, the SLA is probably not yet as effective as it could be. One reason for this is that in many organizations the SLA has been viewed as something that should be avoided at all costs. When it is grudgingly implemented, it is kept to its lowest common denominator. The drive for service level agreements has almost invariably come from the customer side, but the benefits of a well-structured and well-defined SLA are perhaps even greater for the service provider organization.

A service level agreement is a document signed by the service provider and the service recipient, detailing the performance of such service elements as reliability and availability. It has in the past almost invariably been interpreted as a commitment to a certain level of service. Due to the committed nature of the elements within the SLA, the drive from the service provider has been to minimize the scope and usefulness of such agreements. A different approach to the creation, negotiation, and definition of SLAs would enable service providers to reap the majority of the benefits from these documents.

The service level agreement does not have to be something to fear. It also does not have to be a one-sided implementation that places responsibility for service delivery only on the service provider. If an SLA is defined within a service partnership approach, it brings benefits to everyone involved.

7.1 SETTING SLA OBJECTIVES

SLAs come in various shapes and sizes, and there is no template that is ideal for all situations. As a result, one of the most important things to clarify before ever starting the process to define an SLA is what it's intended to achieve. Possible objectives for the service provider organization could be as follows:

1. Setting the expectations of the customer;
2. Setting the expectations within the service provider organization;
3. Providing a means of measuring service quality achievements;
4. Encouraging a service quality culture;
5. Defining what is important to the customer and the supplier.

Once the objectives of the SLA are clear, this provides a framework for defining the elements of service that should be included within it. It can also define the process by which these elements are decided and defined. For example, if an objective of the SLA is to set the expectations of the service recipient, this may determine that the process should be a joint definition activity with the service recipient. The objectives of the SLA determine not only its content, but also how it is created.

7.2 INITIAL DEFINITION OF THE SLA

The initial definition of the service level agreement is perhaps the most important part of the whole process. It builds from the framework provided by the objectives to define the elements of service that should be included to meet these objectives. The commitment nature of the SLAs defined at this stage has normally focused the SLA definition on the number contained within the document. Although there is much discussion of whether it is feasible to commit to 99.97% availability, the discussion often takes place without a clear understanding of what is actually meant by *availability.* Perhaps the parties to the SLA all know they are expecting 99.97%, but 99.97% of what? In these cases an SLA can cause more problems than it solves, because it has probably set different expectations with different groups of people.

By their very nature, SLAs can be an emotional topic and are subject to abuse in certain circumstances. I have come across at least one network supplier whose SLA was willing to commit 100% availability. On further investigation the 100% availability that so impressed their customers turned out to be 100% of zero network throughput. That is, 100% of absolutely nothing—perhaps not as valuable as it at first appeared. To create a worthwhile SLA, take a step back and ensure that unambiguous and well-understood definitions exist for the elements to be outlined within the SLA.

7.3 ENSURING MEASURABLE SLA ELEMENTS

The next important aspect of a service level agreement is to ensure that the elements within it can be measured. Much of the value of an SLA is derived from being able to measure performance and service quality against agreed parameters. It is a good test of whether something should be included within a SLA to ask first whether it can be defined and second whether it can be measured. If the answer to either is no, then further work is required before it can become a valid element within the SLA.

It is also useful to include the measurement methodology within the service level agreement. The way in which certain elements are measured has a direct impact on the level of service that is actually being defined. For example, 99.97% availability annually is a very different target from 99.97% availability measured on a daily basis, as the longer the period the greater the ability to average out any significant service failures.

Measurement also needs to define the perspective from which certain elements are being measured. Whether an application is available is often one of the most difficult elements to measure, since availability differs depending on the perspective of the measurement. An application can be active on a computer but effectively unavailable for effective use. If, for example, the response times are poor or the application is not operating correctly, is the service available or not? Any data processing organization will argue that if the application is running it is available from their perspective. The user will argue that despite the availability of the application in this way, it was unavailable for the user. This is why the definition of terms and the measurement methodology are so important.

7.4 CONTRACTUAL VERSUS ADVISORY SLAs

Such definitions become even more crucial when the SLA is a contractual agreement. One of the reasons behind the reluctance to provide SLAs in the past has been the implicit assumption that they had to represent a contractual commitment. This is not necessarily the case. Many of the objectives outlined above for an SLA can be achieved through the implementation of *advisory* service level agreements. The advisory service level agreement can still perform the functions of setting expectations and encouraging the service culture; although they may not be perceived as being as definitive as contractual SLAs, there is no reason why they should not carry the same weight.

Depending on the organizations involved and the level to which their service partnership has developed, a contractual SLA may not carry any benefit over an advisory SLA. A contractual SLA tends to encourage a minimum risk approach to service commitments, often containing the service levels that can already be achieved. It does not create any drive for improved performance.

The major reason for the contractual nature of traditional SLAs has been a perceived need in the realm of public service providers to offer some form of service rebate capability against poor performance. To date, service rebates in the public domain have been nominal in nature. No public service provider appears to have taken on the challenge of a rebate scheme that includes any form of consequential loss due to the nonperformance of the service—nor would I expect them to do so in the near future. These nominal rebates reinforce the view that the contractual nature of SLAs restricts the scope and usefulness of such agreements without adding any significant value to the process.

A rebate capability does represent a token gesture on the part of the service provider to commit to delivering the levels of service defined within the SLA, and this can be valuable. The introduction of hybrid service level agreements encompassing both contractual and advisory service level agreements represents one way to achieve the scope of the advisory SLA with the token financial penalties behind the contractual service level agreements. This may be the best of both worlds.

The time necessary to define and agree upon a service level agreement within a service provider organization poses an additional problem with the contractual SLA. When contractual and financial implications are involved, the process for ensuring that an SLA can be offered is almost always protracted. This has resulted in a slowdown in the development of the SLA to encompass additional service elements and higher service levels. The advisory SLA, on the other hand, need not be subjected to the same process. Moreover, an SLA should remain meaningful and continue to stretch the service provider organization to deliver a higher quality of service in as many possible areas of the service as possible. The use of an SLA as a driving force for improved service is best achieved through an advisory SLA within a service partnership.

7.5 THE SERVICE DELIVERY AGREEMENT

One of the major benefits of implementing the advisory SLA is its ability to ensure that the service levels being committed to the customer are secured by "back-up" SLAs within the service provider organization. For example, a commitment to the customer is normally between the marketing or product management function and the customer. Any commitment of this kind can be backed up by an SLA between the marketing function and the operations functions delivering the service on a day-to-day basis.

In order to avoid confusing a customer SLA with an internal agreement, the latter can be referred to as *service delivery agreements*. By their very nature they are not contractual, and although they are internal, the same rigorous approach to their definition must be taken for them to adequately support the customer SLA. The development of an SLA with an external customer can be used as the driving force for the development of internal service delivery agreements. This ensures that the commitment to the customer receives maximum visibility throughout an organization and is understood and committed to by all supporting elements within the service provider organization.

The service delivery agreement need not be limited to the elements included within the service level agreement, and it can be used to ensure that common working practices and interfaces across functional units are defined and agreed upon. The paramount objective is to ensure that all groups that contribute to the eventual service being committed to the customer understand what they can expect from other groups and what is expected of them. Through this process the SLA with the customer can be used to bring real benefits to the service provider as well.

7.6 DEVELOPING THE SLA

The development of a service level agreement that is meaningful, measurable, and achievable is a complex process. At each stage of the process, for each element of the SLA, the objectives of the SLA must be met. The process has to be undergone in discrete stages to ensure that the ability of the SLA to meet the objectives is maintained.

7.6.1 Defining Objectives

The first stage of developing an SLA should define the objectives for that SLA. These should be articulated in terms of the objectives of the service provider as well as the service recipient. Possible objectives are as follows:

Customer Objectives

1. To understand the levels of service being offered by the supplier.
2. To feel comfortable that the supplier is able to deliver that level of service.
3. To understand the actions the customer must take to assure those levels of service.
4. To understand how service levels are monitored and measured by the supplier.
5. To understand what additional measurements and monitoring the customer may want to implement.
6. To ensure that the levels of service being offered are appropriate to the customers' use of the service.

Supplier Objectives

1. To ensure that the customer understands the level of service being offered.
2. To ensure that groups responsible for delivering service understand the levels of service being offered.
3. To ensure that the levels of service being offered are actually achievable.
4. To ensure that the levels of service being provided are measured and monitored.
5. To ensure that groups within the supplier organization understand how these service levels are to be maintained.
6. To measure the success of the organization in delivering service.

The above is by no means an exhaustive list of the objectives that can be defined for a service level agreement. They are representative of the scope of the service level agreement needed as a tool for ensuring that expectations and perceptions are correctly set and that an organization can develop a means of measuring their success in providing service against the SLA. The intangible nature of some aspects of service quality does make this very difficult. The SLA is one way of overcoming this problem, at least in part.

7.6.2 Defining SLA Elements

Having defined the objectives for the service level agreement, the scope of the SLA can be established. The scope should be based on the elements that need to be contained within the SLA in order to meet the defined objectives. The elements to be included within the SLA should be those areas of the service that are seen by the customer and the supplier as contributing to meeting their objectives. This will to a large extent depend upon what is considered to be important to the customer and supplier in terms of the service that is delivered. For example, for interactive applications the response time to the user may be a service level element that is required within the SLA. For more batch-oriented data transfer functions it may be data integrity and guaranteed delivery that are important. For customers with low skill level users, the ability of a customer assistance function to answer calls quickly and provide answers immediately will be important. The significant elements will vary depending on the service and the service recipient.

At this stage in the process it is often useful to list all possible service level elements without any restriction based on whether they could eventually fall within the final scope of the service level agreement. Possible elements to be considered could include the following:

1. Application response times;
2. Data integrity;
3. Fault resolution times;
4. Help desk/customer assistance call response times;
5. Maximum number of faults;
6. Frequency of call-back information;
7. Politeness of help desk staff;
8. Availability.

This list is not all-inclusive; the service level elements will depend very much on the development of the objectives and definition of what is important to the service provider and service recipient. Initially these elements will be very loosely defined, but at this stage it is essential to ensure that the elements to be included within the SLA are much more closely defined and that the definitions are agreed upon by the service provider and the service recipient.

The definition at this stage will determine the eventual effectiveness of the SLA when implemented. Availability, for example, is an area in which perceptions are very different. In any networked application, an application can be available while the network is unavailable, thus making the application unavailable from the perspective of a networked user. The definition within a SLA needs to take this into account to ensure that there is clarity in the minds of both the service supplier and the service recipient about when the application is to be considered available.

If an agreed-upon definition cannot be reached, one aspect to consider is whether multiple service elements are actually being defined within a single service element. In the example above, two service level agreement elements of "network availability" and "application availability" could be defined in order to remove the ambiguity behind a single definition of "availability."

7.6.3 Testing for Measurability

Once the definitions are agreed upon, the test of whether the defined elements are measurable within the bounds of the service level agreement can be applied. The process of definition and testing for measurability can be an iterative process, in that the test of measurability can assist in further defining the elements more closely. If an element is not measurable it should not necessarily be discarded, because revisiting the definition can result in the definition of a measurable element.

A distinction needs to be drawn between elements that are not measurable due to the fact that they are subjective in nature and those elements that are not currently measurable due to a lack of monitoring and measurement capability. Service level agreements can be continually enhanced on the availability of new measurement capabilities. An element that cannot be measured at this time may indicate a need to enhance existing measurement tools rather than that the element is inappropriate for inclusion in an SLA.

7.6.4 Setting Service Level Values

When this iterative process is complete, two things should have emerged: a list of defined elements that can be measured and a list of defined elements that cannot currently be measured. The service level elements that are to be offered should now be fully understood throughout the organization, allowing a sensible and informed approach to the definition of the values to be attached to the service level elements. The definition of these values is often the most contentious and difficult part of the process, which is why it is important that the initial groundwork is done to ensure that everyone is talking about exactly the same thing when trying to define the value being attached to a service element. Much of the emotional and contentious nature of the discussion about the values attached to SLA elements is removed when the elements are closely defined. The values defined for the service level elements will also emerge from the objectives originally set for the SLA.

For many service providers the introduction of a service level agreement is one way of trying to ensure that the quality of services is constantly improved. In this case the levels set should be challenging to the organization, ensuring the maintenance of the best levels that are being achieved by the organization. It is, however, important that the levels initially defined are achievable and not merely a wish list for perfection. An SLA can easily lose credibility within a service provider organization if the values are set at a level that is not supported by those who are required to deliver against those service levels.

In the public arena there is a natural conflict within a service provider organization between the groups that want to commit the service levels to the customer and those who are expected to deliver those service levels. The levels within a service level agreement are frequently used as a marketing tool by public service providers in an increasingly competitive market. This is particularly so where customers are evaluating the outsourcing of large mission critical networks or applications. The only way to resolve this is through compromise, but it should be emphasized that a very aggressive service level agreement that is not achievable is of no value to either the customer or the service provider. A service level agreement should above all else be honest in what it presents, as it represents a basis on which a customer can determine the reliance that can be placed on the service and the investment that has to be made in alternative solutions.

If the service is not able to provide high enough levels of service, that is not a fault with the service level agreement, it is a fundamental fault with the service, and no service level agreement can resolve that in isolation. The key to setting the levels within the service level agreement is to ensure that they are challenging, achievable, agreed upon, and honest.

7.6.5 Defining Elements as Contractual or Advisory

When the levels have been set the decision can be made as to which of the elements can be made contractual or advisory. As discussed previously, the value behind contractual service levels is currently restricted due to the limitations on compensation used to support the contractual guarantee.

For service providers the key behind contractual definition is the evaluation of the risk associated with the service level elements being offered on a contractual basis. Again, although there may be a natural desire for a service provider to appear as committed as possible to the levels being offered, the costs—both direct rebate costs and the cost of administering a rebate scheme—of contractual levels of service that are not achieved is significant. The effect on the organization when service is perceived by the employees as contractually unsatisfactory is also an issue. Continually missing SLA objectives, even if they were set unrealistically high, is extremely demotivating for service provider staff. The customer perception of the commitment, despite the rebate, would be as if the commitment to actually deliver those levels of service did not exist. The limited levels of financial compensation will not displace a deeper customer dissatisfaction that has been created by an unachievable service level agreement. It is a fine balance between an SLA that drives an organization to improved service levels and an SLA that is unachievable and lacks credibility.

8 Customer Assistance and Help Desks

Customer assistance comprises the front line troops in service management and the management of the service partnership. It is, in many cases, the image of the service provider most often seen by the service recipient. Yet in most service provider organizations, customer assistance, often referred to as the *help desk,* is the most ignored element of the service. A good guide to a service provider's commitment to the principles of quality service is the relative position of customer assistance within an organization, including the service provider's own perception of the value added by the customer assistance function.

In some organizations the customer assistance function is seen as merely providing a capability to receive calls from users and pass them on, or serving as a gatekeeper between the users and those who can help them. The perception of value of the function within these organizations is very low, as they are effectively made into an enhanced switchboard for calls. The key to resolving customer calls is to get the right information to the right person in the right time frame. In high-volume customer assistance organizations a call dispatch function may be essential to ensure that the information gets to the right person.

Other organizations have implemented highly skilled help desk functions for specific user groups that are able to resolve almost all problems without further reference. The perception of the value of the help desk in the cases I have seen has been so high that some organizations are evaluating the implementation of a second help desk to protect the one that is perceived as high value. Their service recipients will almost certainly perceive this as a retrograde step.

The trade-off between the cost of providing a help desk and the relative value it can add constitutes a constant dilemma. This is normal in any function but is particularly so where the expense of highly skilled staff is high and the majority of calls can be handled fairly quickly by relatively unskilled staff. It is the minority of calls that cannot be handled that cause the problem for the way in which the customer assistance function can be implemented.

8.1 CUSTOMER ASSISTANCE REQUIREMENTS

The requirements for customer assistance differ depending upon the kind of services being offered and also upon the relationship between the service provider and the service recipient.

When multiple services are being offered by a single provider, for example in public telecommunications companies, there has been a strong requirement for these providers to offer a single point of contact for all calls regardless of service. Given the range of services offered it is impossible to provide the highly skilled help desk staff to serve all customers, and some form of call receipt and call dispatch function is almost inevitable with any significant number of users and services.

On the other hand, when a single specialized service is being offered to a small user base it becomes a more reasonable proposition to implement a help desk function that can provide direct assistance to the user for all calls.

The decision of what kind of customer assistance function to implement is key to ongoing customer satisfaction and perception of customer service as well as the cost effectiveness of the service provider organization. The service recipient—as well as the service provider—needs to understand the parameters that determine the provision of customer assistance, because within a service partnership it is possible for the service recipient to help determine the type of customer assistance provided. The following sections examine some of the basic determining parameters.

8.1.1 Number of Users

With a large number of users, there is normally a high volume of customer assistance calls. When call volume is high, the requirement for the quick pick-up of calls and the need to handle a high volume frequently determine the introduction of a call dispatch function. As communications technology has improved, many customer assistance organizations have implemented call queuing systems in order to manage a high call volume with a help desk designed to handle a lower call volume. This enables the customer assistance function to be higher skilled. However, the annoyance of a queued call for a customer with a problem should not be underestimated.

8.1.2 Number of Services

When a number of services are being offered by an organization with a single customer assistance point, it is unlikely that a customer assistance function can provide support for all services without referring the call to a specialist function elsewhere in the service provider organization.

8.1.3 Service Recipient Interface

The type of interface to the service recipient is also a key determinant. Many organizations using public services will implement their own internal help desk to interface to the service provider. This results in two major changes in the relationship. First, the number of calls from a large user population is filtered and reduced, and second, the caller to the service provider customer assistance function is at a higher skill level than the average end user. Having already performed initial problem analysis, the service recipient must have direct contact with someone of equivalent skill level within the provider organization. In these cases a call dispatch function is not appropriate.

8.1.4 Skill Levels

Skill levels of both the service recipient and the service provider are key to the effective working of a customer assistance function. It is at this front line of service delivery that it becomes most clear that service is a dual responsibility requiring a partnership approach.

Within the service recipient a high skill level changes the type and number of calls that are received by the service provider. Problems, for example, will tend to be system- rather than user-oriented (that is, they will tend not to be just user errors). A high skill level within the service recipient also determines the type of response expected from the service provider. When the skill level of the service recipient is higher than that of the service provider, this causes frustration and dissatisfaction in the service recipient. The level of skill in the service recipient can determine the kind of customer assistance implemented by the service provider.

The higher skill levels in the service recipient provide the service provider with the opportunity to increase the skill level within the help desk function, through the reduction in call volume and especially the reduction in calls where the level of skill required to resolve is low. The service recipient does have a responsibility to ensure an adequate skill level exists within their own organization. The calls to customer assistance with problems that turn out to be due to the system not switched on or some other basic lack of knowledge do not help to improve the quality of service. These calls and their frequency force service providers into low-skill first points of contact and a low-value customer assistance organization.

In the long term the development of the skill level within the service provider's help desk is essential to improve the service provided. As the skill levels increase, customer assistance functions are more likely to make an immediate response to users with problems. However, the help desk must be authorized to make the necessary decisions and, at times, system changes to resolve calls. For example, if a software definition for a user is incorrect within a system and requires a change, it is important that these simple changes or fault resolution actions can be taken as quickly as possible without calls being referred from the help desk to a specialist function. The system tools and capabilities required to make these changes are often quite powerful, and unless the skill level of the help desk is sufficient to ensure their correct use, they are also dangerous tools in the hands of an unskilled help desk function.

Service providers are responding to the need for empowerment in a variety of ways, but there is undoubtedly a move to enable front line staff to be able to respond to more and more user needs. This is either through the implementation of help desks with high skill levels or the implementation of better defined procedures to deal with frequent problems. A good help desk today will be able to close about 80% of all telephone calls with the customer on the first call. The objective is to make that as close to 100% as possible.

8.2 RESOLVING PROBLEMS ON THE FIRST CALL

There is a significant amount of effort required to manage common problems so as to resolve them on the first call. The investment in this, as opposed to the resource required to handle a significant number of calls passed through the organization to specialist support, is worth it. The cost of handling any customer calls increases dramatically when multiple support groups within the service provider organization are involved, the complexity of the task of resolving any call increases with the number of people involved, and the possibility of miscommunication or resource bottlenecks increases. All of this detracts from the ability to respond to the customer.

One area that public service providers of low unit value and high volume services, such as domestic telephone services, have found particularly problematic is the area of billing. Billing systems within service providers are invariably complex, and billing queries are difficult to resolve. Resolving the query can often cost the service provider significantly more than the value of the bill being queried. The psychological impact on the customer of being billed incorrectly is also significantly higher than almost any other problem they encounter, both in terms of feeling as if they are being cheated by the service provider and also being unable to trust any element of a bill once one is inaccurate.

From the service provider's perspective, the goal is to resolve these problems quickly and cost effectively, and to change the customer's poor perception of service into a perception of quality service. Help desk empowerment has enabled them to do this. In several telephone companies the help desk or billing inquiry staff are empowered to simply remove a billed item from a customer's bill on the basis of customer complaint, no questions asked.

As a solution to the problem of a customer being incorrectly billed, help desk empowerment is a very powerful way of enhancing the perception of service. For one thing, the atmosphere of trust is reestablished in that the customer feels that they are being trusted to be honest about the incorrect bill. The fact that detailed checking procedures are in place to track abuse of the system after the event is hidden from the customer. Furthermore, it provides an immediate response and, more importantly, an immediate resolution of the customer's problem. The result is a satisfied customer at a minimal cost to the service provider.

The solution also has benefits for the service provider and the help desk. Specifically it enhances the position of the help desk staff as people able to respond to the customers problem. One of the major problems with traditional call dispatch implementations is that the added value of the call dispatch function is minimal and is perceived as such by both the customer and the call dispatch function itself. This will almost always adversely affect the performance of the call dispatch function itself, thereby further reducing any added value that could have been created.

8.3 HELP DESK TOOLS

If the customer assistance function is to be empowered to perform more complex tasks, then the tools to enable them to do this are essential. Recently a lot of emphasis has been placed on the tools and facilities available to the help desk to support the customer. Although these can enhance the capability of the help desk to provide quality service, they are not a replacement for the abilities and attitude of the help desk staff. Like all tools in all trades, if in the wrong hands or used incorrectly, they can cause more harm than good. You can give a bad help desk good tools and not improve customer service in any way at all.

With increasing complexity in systems and communications networks, the tools available to support staff have been enhanced to a remarkable degree. However, help desk tools are not an end in themselves. While it is fascinating to have a network management system that enables the help desk to check the temperature of a processor card in a switch hundreds of miles away, this is probably of little interest to the average help desk caller. The danger of advanced technology in customer facing systems is that the customer assistance organization becomes too embroiled in the technology and less focused on resolving a customer's problem.

The tools used on the help desk need to be the appropriate tools to support the customer. Not every facility available within the system is appropriate for the front line and may detract from the provision of quality service. The identification of the appropriate tools for the help desk is an ongoing task, as the nature of the customer, the customer expectation, and the service itself develops over time.

8.3.1 Help Desk Requirements

The requirements of both the service provider organization and the customer need to be taken into account when the tools are being implemented. The two main customer requirements for the help desk are speed and accuracy. Speed is the ability of the help desk to respond quickly to a call for help and to resolve faults quickly. Accuracy is the requirement that the problem be correctly identified and resolved permanently. The service provider's main requirement is to maintain a high customer perception of quality of service while keeping the cost of the service at acceptable levels.

The above requirements do not necessarily conflict and are in many ways mutually supportive of one another. The requirement for speed and accuracy of response on the part of the service recipient is directly supportive of the desire for cost-effective enhancement of customer perception on the part of the service provider.

A third requirement that a service recipient has of a help desk is less tangible and is measurable only indirectly. It can be described as the help desk's "way of doing things." This is a subjective measurement for the most part, but it can be broken down into elements that give an indication of the performance of the help desk. Within this requirement I would include elements such as politeness and general helpfulness of the help desk function. Although less tangible than speed or accuracy, these aspects of customer support are a determining factor in the perception of quality service.

My worst experiences with help desks have almost always been based either on being treated like an idiot or on being made to feel as if my particular problem was of no real importance. There have also been occasions when the help desk has been fundamentally inept at resolving a problem but has clearly tried and has been genuinely concerned at the failure to resolve the problem. In this case, contact with the help desk ended with an unresolved problem but a customer who was satisfied with the service he had received.

The difficulty for the help desk function is that they only receive a call from a user who is experiencing difficulty. This user is probably frustrated, angry, and unable to achieve their own objectives due to a perceived failure in the system. This is probably not the best psychological starting point for a conversation with someone who is impolite and unhelpful.

Tools need to support all these elements of the help desk requirements. When a manual system or support without the use of system tools can meet these requirements there is little need to develop automated facilities to meet the same need. Help desk support tools are available today to support every stage of the help desk process from call receipt to detailed system facilities to aid problem diagnosis and resolution. The right tools for the right job will depend upon the services being offered and the expectation of service recipients.

8.3.2 Initial Help Desk Contact

Probably the biggest initial impact on a frustrated and angry user needing help is the delay in obtaining help. It is likely that prior to calling a help desk the user has tried several bypasses to the problem, talked to at least two "experts" in the office, and then decided to call the help desk as a last resort. From this point on, the help desk has inherited a problem with the service and several other problems generated for them. By this time, the user has already decided the service is unusable and understood by nobody. The help desk has to turn that perception around.

Although the user has probably spent a good period of time prior to calling the help desk, the problem is now urgent, and every second counts. Problem resolution has already been delayed by minutes if not hours. However, the first delay in resolving the problem that the user will acknowledge is any delay in picking up the call by the help desk. All aspects of call pick-up and initial contact are some of the most important in determining the eventual customer satisfaction with the help desk function. Call management systems can help this process in several ways.

Most systems will provide the ability to monitor call pick-up delay. It is essential to equip the help desk sufficiently to be able to meet defined call pick-up targets. The time taken to answer the telephone is beginning to be defined within service level agreements at a target performance level—normally between three and five rings.

Call management can also provide details on the number, length, and distribution of calls throughout the day. These are important details for determining help desk call answering resourcing. Given the normal state of mind of a help desk caller it is probably unwise to implement taped "your call has been queued" answers in this environment, as they are likely to make the help desk's job even more difficult.

8.3.3 Fault Logging

From the initial call answering, the next stage in the process is the actual logging of the fault details. Fault logging procedures and systems again have the ability to irritate a help desk caller beyond the bounds of human tolerance. In order for a fault to be analyzed and resolved, a certain amount of information is required from the caller. However, there are systems that are rigid in the definition of the information required. Much of this information will appear to be irrelevant to the caller (as undoubtedly some of the information will be for the internal purposes of the service provider) and unrelated to the problem the caller needs resolved. Fault logging systems need to be able to avoid this pitfall, as once again it sets a poor perception with the customer and creates antagonism toward the help desk operator.

In many environments it is unlikely that the user calling will have any administrative information required for service provider internal purposes. Long conversations in which users are asked to provide reference numbers, account codes, system type details, and other extraneous information should be avoided if at all possible. It just adds to the frustration.

8.3.4 Problem Diagnosis

Once the fault has been logged, the call moves on to problem diagnosis. Normally specific to the service or systems involved, problem diagnosis and resolution tools may need to be available to the help desk. It is often in this area where the conflict between the level of educational and skills investment in the help desk emerges. System tools that can enable the help desk to provide a higher level of service to the customer also carry with them a certain amount of risk in their potential misuse. There is the possibility of both accidental and deliberate and malicious system damage. There is no right or wrong answer to how many system-oriented capabilities should be made available at the front line; it depends on the kind of function you want the help desk to perform.

The level of risk involved from accidental misuse is minimized by education and good procedures. The dangers of deliberate misuse emphasize the fact that every area of the service provider organization needs to be directed toward delivery of service to the customer. Clearly recruitment and personnel policies have a significant role to play in minimizing the risk of misuse.

If the first-line call handling and call dispatch function has been implemented, in many cases the system tools required are minimal. If, on the other hand, it is to be a full service able to resolve problems without needing to hand them off to more skilled "back-room" support groups, then the help desk requires access to all the system tools available.

Speed of response is faster if the help desk can resolve the call without further reference, but the cost of the help desk increases as it becomes a higher skill base. Greater education and training costs are incurred to manage the higher number of tools. It is also normally the case that there are a greater number of help desk staff than higher level specialist support staff, increasing the risk associated with enabling a large number of people to have access to restricted system utilities. All these factors need to be taken into account when determining the tools to be made available.

There are, of course, coordination and call tracking costs incurred when the alternative implementation approach is adopted. When a problem has to be handed off to other support groups within the service provider organization or in subcontractor organizations, the problem needs to be tracked against set target times for responses. The help desk is normally the best place for this tracking to be performed and for primary ownership of the problem to reside for interface to the customer. At all times, the customer interface must have the information available to update a customer on the status of a call or problem that is being handled.

Once a problem has been handed off to the technical support areas, the help desk can find themselves having to continually chase the status of a problem internally. This not only consumes a lot of time and effort, but also creates a very poor customer perception. When the problem resolution responsibility is separated from the customer facing groups the level of customer focus can be diminished. This is one of the intangible service costs of using "back-room" skilled support who do not work directly with the service recipient.

8.3.5 Problem Closure

As important as call receipt and problem analysis are systems to enable the problem to be closed. When a problem is closed, both the help desk and the user who called the help desk should both agree that the problem is resolved. There is normally some understandable pressure to close problems within a set time frame. Without this check in place I have seen large numbers of problems closed, only to be replaced by a new problem when the user still believes the problem exists. Technically a problem can be resolved without the user seeing the resolution. This could be, for example, because the system requires rebooting overnight for a change to become active. In these cases the service provider may believe the problem is resolved but the user would not see it as resolved until the next morning. In order to retain an accurate view of how problem resolutions are perceived by the customer, the problem should not be considered closed until the user agrees it is resolved. Problem management systems and procedures should also take into account the recording of how the problem was closed in order to enable future occurrences to be resolved quickly and efficiently.

8.4 HELP DESK INTERFACE TO CUSTOMERS

The help desk represents a service provider to its customers and is often the main point of regular contact between the two organizations. Particular importance needs to be placed on the way in which this interface operates. The point of contact tend to vary depending on both the organizations and the services involved. There are two main methods of providing this link with respect to public services that are being used.

8.4.1 Internal Help Desks

What is becoming the prime method is for customer organizations to manage the contact between their staff and the supplier organization by implementing an internal help desk that will manage the process of dealing with the service provider. This has become prevalent in areas where, for example, basic networking capabilities are being provided by the service provider, and the business applications are being operated by an internal systems organization. In these cases the majority of user problems tend to be related to the internal applications. On the basis of trying to resolve the majority of problems as close to the user as possible, it makes sense to have an internal help desk take initial calls and deal remotely with the network provider.

The disadvantage associated with this approach is that the resolution of networking problems requires an interim step before the user's problem reaches the network service provider. However, because the analysis of the problem (to the extent that application-related problems can be ignored) has already been completed, the problem management process may be able to operate more effectively.

The disadvantages of the interim step can be minimized if the help desk tools and capabilities are made available in some form to the internal help desk for all aspects of the service, including that provided externally. If the external and internal help desks can operate from the same problem management and tracking systems, the effectiveness of the interface between them is greatly improved. When a complete problem history can be shared, more information is available to assist in resolution, even if only to see which possible causes have already been eliminated. When the help desk group closest to the end user can see the current resolution status and activity in both organizations, the service provided to the end user can be improved.

8.4.2 Direct Contact with the Help Desk

The alternative to an internal help desk is to enable all users to contact the service provider's help desk directly. This approach has disadvantages for both the service provider and the customer organization in terms of the number of communication paths between the two organizations and the resolution of internal problems within the customer environment. If all problems are addressed to the external help desk, then a significant number will undoubtedly be referred back to the customer as problems with their application or technical environment. This is inherently less efficient and tends to be a more cumbersome process than the filtering of calls before they leave the service recipient organization.

Again, there is no right or wrong answer as to the best interface method. It has to be based on the specific applications and systems being supported. The guiding principle, however, should be to try and resolve as many problems as possible as close to the user as possible.

It should also be remembered that the help desk interface is not a one-way relationship. The help desk is as dependent on the user as the user is on the help desk. If a problem is to be resolved quickly, the users must be able to interpret symptoms they see and report them in such a way as to aid the problem analysis and resolution performed by the help desk. A user population that is poorly educated and unable to provide this level of support to the help desk can be a costly population to support.

8.5 PROACTIVE HELP DESK ACTIVITY

One of the reasons why so much emphasis is placed on calls to the help desk is because traditionally the help desk has been a reactive organization, responding to customer calls and acting on customer requests. However, a quiet customer is not necessarily a satisfied customer; in some circumstances, if customer satisfaction is to be maintained, the help desk needs to become a more proactive function within the customer support organization.

Specific examples of proactive implementations can now be found within several public service provider organizations. They include proactive status reporting on problem resolution activity at preset time intervals and preemptive notification to customers of systemwide user-affecting problems. These initiatives can be very valuable in maintaining customer satisfaction and developing the perception of a competent service provider organization. If the help desk telephones you to tell you there is a problem before you have noticed one, this creates a positive impression. From the service provider's point of view, it also avoids the nightmare of every user calling the help desk simultaneously.

The key to proactive help desk activity is the alert systems built in to the services being provided. There tends to be an assumption that when a message about a problem is displayed on a help desk monitoring screen, someone will act on it. The chances are that, due to the reactive nature of most help desk activity, nobody is even monitoring that screen on a regular basis. Alert capabilities need to be specifically tailored to ensure that action is taken.

Proactive monitoring and reporting by the help desk can reduce costs through the reduction of incoming call volumes. If the help desk is able to let people know that a problem exists and is being resolved it stops people from calling the help desk. It is normally easier to manage making calls out of the help desk than it is to manage a large volume of incoming calls. Again, the integration of help desk support tools and systems with the systems within the service recipient organization enables a more proactive approach to be taken.

The help desk is a complex environment to manage. It deals with dissatisfied users on a regular basis—nobody ever telephones a help desk to let them know that everything is working well. However, the help desk has the ability to change a user's perception of the service for better or worse every time an operator picks up a telephone. This is what makes the help desk so important—as an organization of ambassadors for the service and for the service provider organization.

9 Service Reporting

The monitoring and measurement of service performance, followed by the reporting of performance, make up one of the most important elements of service management. Unfortunately, because it tends to be a fairly unglamorous activity and is not an exciting new gadget that the user or system developer can play with, service reporting often does not receive the attention it deserves. Frequently services are implemented when the user function is fully available, and it is considered acceptable to implement the service without the ability to monitor, measure, or report on service performance. However, it is impossible to accurately gauge the quality of the service being provided without having the information available that measures service performance. This is a like flying an airplane without an altimeter—very brave, but very stupid. Intuitive feelings about the quality of the service are not enough to manage the service effectively.

The three activities of measuring, monitoring, and reporting are very different functions, each of which is essential in its own right, but only when the three are integrated and seen as a whole can they really be used as a tool for service management. *Monitor-*

ing is the collection of raw information concerning all aspects of service performance. This provides the basic information that is required for measurement and reporting activities. *Measuring* is the use of the raw information to evaluate service performance against agreed tolerances. *Reporting* is the dissemination of the information gathered and developed through monitoring and measuring processes.

The monitoring and measurement of the service form the base from which to show the performance of the service, and the reporting elements of the process provide the ability to manage perceptions of the service performance. Monitoring and measurement are really supporting functions for the eventual reporting of service performance, but they also have value in their own right. The objectives of the whole process are to ensure the provision of high-quality service and to ensure that the quality of service being provided is understood.

It is possible to monitor, measure, and report on everything associated with a service, but more is not necessarily better in this case. The value of the capability is in the action that it will generate. A beautiful reporting system is not the end in itself, it is a tool for service management. As such, and as a method for influencing perceptions of the service and levels of satisfaction, the capability must retain credibility with all those who use it. A reporting system that gives a clear picture of useless information about the service will soon lose all value and no longer be an effective service management tool. In order to make this tool as effective as possible it has to be closely targeted on specific elements of the service that determine the quality of service being provided. Defining those areas to be monitored, measured, and reported is worth investing some time in, as it is the base on which the rest of the capability is built.

Keep in mind that the three elements of the process are very closely linked. It is not possible to report on the elements of service that are not measured, nor is it possible to measure elements of service that aren't monitored. The minimum elements to be covered within the process are those that are to be reported on, but additional areas may be measured and monitored without necessarily being reported.

The first aspects of the service that need to be covered within the process are those aspects that are committed as part of a service level agreement. There is little point in having a service level agreement if it isn't possible to show whether the goals of the service level agreement are being achieved. If the service level agreement has been defined correctly then this provides a good basis for the reporting system. The key elements of the service level agreement will have been defined as being measurable, and also being those service elements that are actually important to the service recipient. If these two criteria have been established as part of the development of a service level agreement, then they provide a very good starting point for the definition of service elements against which performance can be reported.

9.1 KEY PRINCIPLES OF REPORTING

The reporting of the service is one of the main aspects of managing customer perceptions of the service, and thus satisfaction. The credibility of the reporting has to be maintained in order to achieve this, and there are several key principles that should apply to all reports produced.

First and foremost, any reporting of service performance must be accurate. An inaccurate report will soon be relegated to the trash and will not only fail to improve customer satisfaction but will actually damage satisfaction levels. At best it will be interpreted as ineptitude, at worst as dishonesty on the part of the service provider.

Second, the reports have to be honest. It is unlikely that many service providers would contemplate reporting untruths, but any statistically based report can represent the truth in such a way that it is dishonest in all but name. A service recipient will soon see this for exactly what it is, and again customer satisfaction will be damaged. If service performance has been poor, chances are that the service recipient has noticed this anyway, so there is little to be gained by not admitting it. If anything, well- managed contrition on the part of the service provider can have a positive impact on customer satisfaction.

Third, the information reported has to be relevant to the people who receive the reports. While the maximum temperature of a card in a switch may be of great interest and importance to those responsible for maintenance of the switch housing environment, it is of little interest to anyone else. Again, sending people the wrong kind of information is sufficient to destroy the credibility of all the information sent and damaging to customer satisfaction.

Ideally the information would be presented to show the performance of the service in relation to the service recipient's reasons for using the service and the importance of the service parameters to their business. Users of an electronic mail service, for example, may be fascinated to know the average message routing lookup table delay for mail items, but the report that would be of real value would reveal the length of time for complete delivery of a message to the intended recipient.

Assuming that accurate and honest information has reached the right people, the fourth requirement is that it is understandable when it is received. If information is not presented clearly it is either ignored or interpreted as being an attempt to hide the truth.

Fifth and finally, the information has to be available at the right time. This applies to all reporting activity regarding service performance. If the information indicates tat actions need to be taken then it must be available in time for these actions to be implemented. If it is just information to report on performance, then in order to manage customer perceptions it needs to be available in a time scale that allows it to have an effect on customer perception. If the reports are available too late then it can either reopen old wounds or have no impact at all.

9.2 THE APPROPRIATENESS OF REPORTING

The above basic principles provide a starting point for determining the reporting structure both internally and externally. The main questions to determine the appropriateness of reporting can be based around these principles to make reporting effective. Perhaps the best test for reports is to ask why the information is being reported. This question leads to two major subsidiary questions that should have answers before reporting processes are implemented.

First, who is the report intended for? Defining the audience establishes the levels of basic knowledge that can be assumed for the recipient of the report and can determine the relevant information that should be included. There are many different audiences for reports generated to support service management, within both the service provider and the service recipient. All of them are equally important because all are linked through the provision of services. It is essential that all parties understand the current level of service performance and capability so that they can understand how each perceives the service. The reporting process is an essential part of managing the relationship between service provider and service recipient and keeping that relationship working on the basis of a mutual understanding of the service.

The second question is, what are they going to do with the information? The information required within a report is also determined by what it is to be used for. A report that gives information to determine investment decisions would contain very different information and analysis than reports required to manage system capacity for example.

9.3 MEASUREMENT AND MONITORING

Measurement and monitoring are essential elements of managing service performance because they are often the first indicators of potential service problems. While reporting represents a means of managing customer perception of the service, it is more effective to use monitoring and measurement tools to manage service performance proactively. Much of the capabilities for monitoring and measurement are now automated, and the use of intelligent network management tools for proactive service monitoring and problem resolution paves the way for higher levels of availability and performance.

Proactive service management requires monitoring tools to be effective in tracking elements of the system that indicate possible problems emerging. However, although much has been automated, the monitoring of performance is only an enabling tool; much of the responsibility for ensuring that identified possible problems are resolved rests upon the actions of service support staff.

Similar principles apply to monitoring and measuring as have been outlined for reporting. The information presented has to be available to support staff in the right format and at the right time. The use of graphical user interfaces has now provided capabilities for the early identification of problems that were previously not feasible with text-driven alert systems that produced thousands of lines of warning messages that operators needed to judge as important or not. In the same way that the credibility of reporting systems depends on the information being relevant, a monitoring system that is constantly producing false alarms will eventually ensure that the real problems are ignored.

There has always been a tendency to monitor and measure everything just in case it was needed, but as with reporting systems, this detracts from the relevance of the information. It also adds a degree of overhead to any system that is often not justified. The question of why something is being monitored and what actions will be taken as a result of it being monitored is a good measure of the appropriate level of monitoring tools to be implemented. In particular, how it will improve service performance should be the cornerstone on which service monitoring decisions are based.

9.4 CONCLUSION

The monitoring, measuring, and reporting capabilities of a service provider are often a key indicator to just how seriously they are managing the service they are providing. These functions can also indicate whether the service provider is just hoping that the service is alright and relying on customer complaints to indicate otherwise. The proactive management of service performance relies upon having clear advance warning of service performance issues and reporting capabilities that enable trend analysis and action to avoid service problems. Without these tools service managers are flying blind, and it is possible that the whole service is at risk. Without a clear understanding on both sides of the service partnership of the levels of measurable service being delivered, perceptions of the service will differ.

Creating common perceptions and a common understanding can aid the development of a working service partnership. As a tool to enable service management, the right service performance information is absolutely essential. It can provide the early warning necessary to avert problems with the service or with perception of the service and the information necessary to improve service levels. It can support the development and management of service level agreements and assist in ensuring resources are correctly deployed. In a complex service environment there are many parties involved in either the provision or receipt of service, and conflicts invariably arise. Without the right tools to support the management of these conflicts, they may never be resolved.

Conflicting Service Requirements 10

One of the most fundamental aspects about service provision is that it is essentially about relationships—between people or between people and systems. The help desk has a relationship with the user; the customers have a relationship with the service provider. Service provision and receipt is an ongoing daily relationship; everything that is done within that relationship will have some impact on the other parties. It would be relatively simple if the relationship was between one homogeneous unit called the service provider and another homogeneous unit called the service recipient. Unfortunately it isn't. There are many different parties involved on both sides, in a web of relationships that can be very complex to understand and difficult to manage, especially since not all the elements of the relationship have the same objectives.

10.1 CONFLICTS BETWEEN SERVICE RECIPIENTS

This situation is most obvious in the provision of shared public services, but it is equally applicable to all types of service provision, even if the scale of the problem is not so great. In shared services there are a multitude of service recipients all in a relationship with the service provider. As a result each of the service recipients has an indirect relationship with each other, in that they are all bound by use of the same service. Immediately there is a potential for conflict. Each of these service recipients are effectively competing for the resources of the service provider. They may all have different uses of the service and different priorities for what they expect from the service provider. It is the job of service management to ensure that these relationships are not allowed to conflict significantly with one another.

A simple example is one in which a shared host computer environment is being provided for several different communities of users. One user can normally perform tasks that will degrade the service available to other users, and the management of the service should ensure that this is not a common occurrence. Although the problem will probably be seen as belonging to the service provider, it is in fact the indirect impact of a conflict between service recipients.

Rarely is there a malicious attempt to degrade services provided to other users, but network viruses are an example of this phenomenon. An individual user can maliciously generate a virus that will consume all available network resource, making the network unusable for all other users.

User groups and common interest groups have been formed by service recipients to deal as a single entity with service providers, but these have tended to focus on the relationship between the service provider and the service recipient rather on than conflicts that exist between each of the service recipients. Ideally these groups would also focus on the conflicts between each other and how their indirect relationship is managed through the service provider.

10.2 CONFLICTS BETWEEN SERVICE PROVIDERS

In addition to the conflict between service recipients, there are also obvious conflicts between different service providers. It is rare that a single user organization has only one service provider for technology-based services. More likely, the service providers have an indirect relationship through a single service recipient. Many user organizations find major problems in getting two suppliers to work together to provide a complete solution. Conflicts between service providers are one of the single biggest obstacles to progress for the service recipient with multiple suppliers, and it is normally the service recipient who has to manage this conflict.

It is difficult to persuade potentially competing providers to work together. There are a variety of issues that make this problematic, for example, who owns the customer relationship, and how the disclosure of proprietary information is managed. Unless there is a complete change of attitude on the part of service providers, the service recipient needs to maintain very strong logical barriers between their service providers. This in turn prevents any multilateral service partnership from ever being implemented—or, if implemented, from ever being effective.

10.3 CONFLICTS WITHIN THE SERVICE PROVIDER

There can also be conflicts within one organization, particularly the service provider. Within service provider organizations there is naturally a constant battle between the cost of providing services and the quality of the service being provided. In the past, the lack of focus on the intangible elements of the service has sometimes caused the service delivery function to be a Cinderella to the service provider organization—constantly underfunded and understaffed, with the majority of expenditure focused on new function and features for the technical base of the service being provided.

10.4 MANAGING CONFLICTS

The service provider may find that the variety of demands from users makes it almost impossible to maintain the quality of service required by all the users because their requirements are so different.

This may be the result of either the service being oversold to the customers or the expectations of the customer not being set correctly regarding all the aspects of the service.

In some ways, this issue represents everything that service management is trying to achieve—to ensure that the conflicts inherent in the service relationship are understood and that they do not affect the perception of the service. The concept of treating the service provider and service recipient relationship as a partnership is to ensure that these conflicts are managed together rather than in opposition to one another. Service managers need to ensure that conflicts are understood by all parties to the relationship—including indirect conflicts between different service recipients. Some of the specific areas to be clearly defined are discussed below.

10.4.1 Performance Impact of a Single User on Other Users

In most shared services this is unavoidable, and there is little that can be done to manage it once it happens. The proactive management of this situation is essential. The only reactive management that I have seen that is effective is the removal of service access form the offending user. This is a somewhat utilitarian view of service provision, in ensuring the greatest good for the greatest number, but the users for whom access was denied certainly did not consider it an acceptable solution!

Service performance limitations need to be clearly defined and understood. If what the users are trying to do with the service does cause problems, then they are probably using the wrong service, and an alternative solution needs to be found. This can happen as a natural development; for example, almost all systems or services originally designed for electronic mail, which was thought to consist only of very small personal messages, are now suffering from being used to send large documents or spreadsheets, thereby causing service performance issues. It is ironic that in these cases services are victims of their own success.

10.4.2 Enhancement and Stability

Most users need a service to develop with their business needs and change with them as their requirements change. This is in conflict with those users for whom the service is adequate; they need a reliable and stable service. Almost any change or enhancement to a service will affect reliability for a period of time after the enhancement is installed.

10.4.3 Cost and Quality

For some reason it is no longer acceptable to suggest that there is a cost and quality conflict, however it is probably the most obvious conflict in the service environment. The quality of people servicing the customer is directly proportionate to their cost, and service remains a people-intensive business regardless of technical advances.

To manage these conflicts the service provider must ensure that the whole service community can reconcile the conflicts with their own objectives and the objectives of others who are brought together through their use of the service. Constant management by user groups of the user community is required to ensure that their objectives for the service are not completely opposed to all other members of the service partnership. It is a key achievement of service management to maintain the capability of all service users to act as a community with overriding common goals that make them a part of the service.

11 Implementing Quality Services

It is a fact of life that it is very easy to implement a poor quality service. It is then very difficult to take a service that was implemented poorly, and through a process of improvement, change the user perception of the service. In services, first impressions last a long time, and bad first impressions last even longer than good ones.

For this reason, the implementation of a service is one of the most important elements of the whole service life cycle. Although service implementation can be used to describe the whole process from gathering requirements from users to managing the use of the service, in this chapter I would like to focus on the period in which the service recipient and the service provider have agreed on the service that is to be offered, the system elements have been developed, and the focus of the service is the implementation of the user environment and bringing the user population on-line. This particular aspect of service implementation is becoming increasingly important as the demands for flexible and rapidly changing services accelerate.

11.1 STEPS TOWARD ACHIEVING QUALITY IMPLEMENTATION

As previously mentioned, change is also a major cause of service problems. Any service implementation is fighting a hard battle to ensure timely implementation with no adverse user impacts. There are two main steps in service implementation that can help to achieve that.

11.1.1 Implementation Planning

Implementation planning is the cornerstone of an implementation that will create customer satisfaction. First, it establishes the expectations of the customer regarding the time table involved in the implementation and, second, it should indicate all the tasks and the interdependencies that will need to be completed if the implementation is to be successful.

Implementation planning is becoming more important as technology is being used to create competitive advantage. Businesses need to plan around when a new service feature will be available. The importance of timely implementation is reflected in public services by the availability of penalty payments from suppliers for the late implementation of services—regardless of any good reasons.

11.1.2 Service Testing Process

The second key element of an implementation is to ensure that when it is deemed to be available it really is available to fulfill the purpose for which it was intended. It is often accepted that the first few weeks of a new service, or in some cases even the first version of a service, will be somewhat unstable and not for the faint hearted. Suppliers even have a term for customers who take the first version of a service—the "lunatic fringe"—implying that they really should know better than to expect it to work. With many businesses now dependent on technology to survive this should no longer be acceptable to anybody.

11.2 PROBLEMS WITH TESTING SERVICES

There are three problems with testing services. First, it is terribly boring for those who have to do the actual testing. Second, testing is normally done very late in the development cycle of a service, so dates for availability have already been committed. The availability of the service with a few "rough edges" can become an option in order to meet the date. Finally, the testing of the intangible elements of service provision is not normally adequately addressed. In many ways all of the problems are related. They are worth overcoming, as it is the testing that finally determines the quality of the first impression given to the customer.

It is easy to imagine the nightmare implementation process for a new service. A traditional service development and implementation process will involve several stages of testing, conducted by different groups of people. First there is a unit test, in which individual elements of the service such as programs or pieces of hardware are tested by those who installed them. This is normally done in isolation and is usually not well controlled. There is a strong dependency on the developers' sense of fair play and personal pride to ensure the adequacy of testing. The unit test is followed by some kind of integration test, in which the individual components of the service are finally put together in something approaching their service implementation design. The only problems at this stage are that half the components are not yet developed, the sense of fair play deserted the developers of the other half, and the available components don't work.

By the end of the integration test the developers have had another good idea and have given up on these minor implementation problems. In their mind the service is ready for a full test by the independent test group. This test group has had little or no education on the new service and has certainly had most of the existing problems hidden from them by the developers. They spend the first week in shock!

The customers by this stage have probably been given a date by which they can expect to have a fully functional service. After the independent test has been completed, a traditional approach often will then include a "beta test" with customers testing the service.

Of course at this stage the independent test group have given up testing things due to the large backlog of unresolved problems. The customer, who has probably got better things to do, tries to test for half a day and, having failed to get anything to work at all, decides not to bother. Having not had a complaint from the customer, the service provider decides the service is perfect and implements it immediately. This time everyone gets to spend the first week of shock.

All of the above testing process is riddled with problems that the test group are miraculously supposed to be able to counteract to ensure perfection by the service reaches the customer. But of course nobody really implements services this way—or do they?

11.3 RESOLVING PROBLEMS

It is essential that the majority of problems are resolved early on in the development cycle while the service developers are still closely involved with the implementation and the impact of changes to the service is minimized. It is cheaper and easier to resolve problems at this stage. It also ensures that later in the testing process the focus can be maintained on the intangible service testing rather then on relatively straightforward functional problems.

The demotivating effect of finding a significant number of functional errors late in the testing cycle is enormous. The problems are much more visible to all concerned, including those who will be responsible for supporting the service on a day-to-day basis. If their first impression is of a problematic service, this will immediately generate a negative view of the service, which is then transmitted to the customer.

What is wrong with the normal testing process is that it enables problems to be avoided or delayed until the final stages of testing. This forces all issues to a period of time in service implementation when they are most costly and have the most negative impact on the service. There are some ways in which these problems can be countered, although I cannot claim to have seen a complete solution anywhere.

By not leaving the independent testing to the end of the cycle it is possible to identify a greater number of problems early on. The testers who would normally be involved in the final service testing could be involved in integration or even unit testing. In addition to this, it is possible to make the developers of the service (who would normally be working on other things) work on the final implementation testing of the service.

Finally, the concept of acceptance can help to ensure that a system passes from one stage of development to another in a quality manner. The function that is completing the next stage of the process should formally define the requirements that they want to see completed in order to accept the system for their stage of the process. This can be implemented between each of the defined stages of testing, ensuring that criteria are met from exiting one stage and entering another.

A certain amount of discipline and courage is required for acceptance to be refused. It is superficially easier to assume that problems will be fixed in the next stage of the process, but it is often a flawed assumption and always detracts from what the next stage should really be trying to achieve.

11.4 CUSTOMER BETA TEST

The final stage of any testing of a service should be testing the service intangibles that are so important to the customer. In some ways it is possible to look on this as testing the *people* who are going to provide the service on a daily basis—and I think this is one of the reasons why it is so infrequently done. There is, in my experience, more than enough emotion involved in telling a programmer that his work of art doesn't actually do what it was supposed to. It is even more difficult when it is not the code that is at fault but the service provided by an individual.

However, it is essential that they be tested, and the usual way of doing this is by trying to involve the customer in a beta test or pilot of the service in a live environment. This is probably the only way in which the concept of testing the people involved in service provision can be managed effectively without destroying working relationships within the service provider organization. Unfortunately the effective use of test customers is a black art in itself, and the whole area of beta tests, trials, and pilots needs very careful management.

Preproduction tests with customers can be very valuable, but they can also take up a lot of resources without achieving very much. In order to get the best out of them they need to be taken very seriously and not implemented as a last-minute activity to iron out the remaining bugs.

Customers to be involved in beta testing should be identified very early in the development and implementation cycle. There are two reasons for this. First, the customer could add significant value to the whole testing process and not just the final stages when it is often too late to change anything. Second, it is very difficult to in a short time scale for the customer's participation in a beta test to be productive. Ideally the service provider would want any beta test customer to start using the service from the beginning of the beta test period, but there is a whole process of education and preparation for the beta test that needs to be undertaken before any real testing can take place. Test preparation in this way can make the test worthwhile. If the customer starts testing with no clear idea of why they are involved in the test, it is unlikely that any real value will be gained by either party.

Prior to any test, the service provider and service recipient should spend time together to:

1. Agree to the test objectives.
2. Agree to the criteria that determine whether the test is successful.
3. Agree to any specific areas of testing to be undertaken.
4. Define the test time scales.
5. Agree to and commit the resources required to complete the test.

Ideally the best test environment would be exactly as the service will be when running live, but it is a brave service provider that doesn't add additional specialized support resource much closer to the customer than is normal. This is normally assumed to be way to ensure that everyone gets the best out of the test, but it may actually be self-defeating. Providing additional resource at this stage immediately distorts the test environment and negates the findings from the test associated with any area where specific support has been implemented. I suspect that there is no right answer, and it will depend upon the customer involved and the objectives for the beta test.

It is imperative not to overlook the definition of what is a successful outcome of the test. This obviously varies according to the service and will probably be different for each customer involved. One of the problems with beta tests is that they invariably seem to assume that there is a successful outcome. I have personally never come across a service that was withdrawn as a result of a beta test "failure." This may not be a problem, but it does question whether there is any point to beta testing at all.

11.5 CONCLUSION

The problems and issues with service testing too often lead to systems being implemented after insufficient and incorrect testing having been performed. Function probably does get adequately tested through the process, but far too often services become available to users when the help desk has never been tested, the support procedures have never been exercised, and reporting and monitoring systems have never been reported or monitored. Sometimes this may represent an acceptable risk for all concerned, but too often at least one side of the service partnership is under the impression that all these issues have been resolved.

All these issues serve to create the biggest problem which is that the testing process is no longer seen by many people within service provider organizations as adding sufficient value to justify what is a very labor- and cost-intensive exercise. The fact that problems are not found at the right testing stage causes dissatisfaction with the next stage when they are finally found. The continued testing of function rather than service leads to the same tests being performed again and again. The poor management and lack of value of beta tests leads to dissatisfaction with these initiatives and all in all creates a complete testing process that is worthless in many cases. No wonder the first few live days of a new service is a testing time in every sense of the word.

The handover from an internal testing environment to a full-service environment is an extremely difficult process. It is a firm law that all systems that operated successfully on the last day of testing fail to do so for previously unknown reasons on the first day of live service. This law exists only because the test environments and processes that have grown from the days of testing software developments are basically inadequate for testing a service, and because existing processes have failed the service recipient by driving a "let's test it in the real world" approach to service implementation.

The above picture is the nightmare every service implementation has to avoid. By defining objectives and ensuring that each planned stage of testing achieves them, it is possible to avoid many of these problems. Most help desk and support staff that I know would welcome an opportunity to be much more closely involved with service development and implementation to ensure they can deliver quality service when they have to. Certainly all customers would welcome a service that works from the first day of implementation.

Service testing has to be acknowledged to be fundamentally different from any other test requirement. The need to test the processes, the skills, and the people is different from a normal product development requirement. Many service providers are at last starting to understand that the testing process adds value to the implementation of service. Customers are driving the beta test environment to a state where it adds value to the whole service partnership. Perhaps one day there will be no need for the "lunatic fringe."

12 Doing It Again Tomorrow

Service management is a task that never ends. Every moment of every day creates new situations and new challenges. They all require a creative and imaginative response to ensure that customer satisfaction is maintained and service quality levels do not fall. This is what makes it interesting, but it's also what makes it difficult.

The traditional response to performing repeated tasks to a consistent level of quality is to document and proceduralize the responses to situations in order to maintain that level of quality. This forms the basis of certification to recognized international quality standards such as ISO9000. However, in a service environment, although it is possible and desirable to proceduralize many aspects of service provision, procedures alone do not in any way guarantee customer satisfaction or high-quality service. Service management is more than the procedures.

A procedure-driven service organization can ensure that it achieves an outstanding level of mediocrity though the implementation of good procedures that are maintained and adhered to religiously. The true test of the service organization and service management is the ability to react to the occurrences that are outside the bounds of procedures and processes. In the same way that the measure of customer satisfaction is determined by the perceptions of individuals, so the continued provision of service at levels that will achieve customer satisfaction is determined by individuals, this time within the service provider organization.

However, the individuals must be working together as a team. The quality of service received from contact with any particular individual should not exceed all others within that team, as this creates an imbalance in customer expectation. There are occasions when customers, for example when dealing with a help desk function, will insist on dealing with a specific individual who has worked well for them on previous occasions. This has both positive and negative effects. It does build a service relationship and enable a deeper relationship to be built and maintained, but it can also cause overload of specific members of the help desk function. If one individual is providing excellent service, this should become the benchmark for all members of the service management team.

The motivation, training, and personal values of the individuals within the service provider organization are essential aspects of service management. To ensure that the organization can respond to both ordinary and extraordinary occurrences, the individuals within the organization must be able to respond to them. Moreover, they must be motivated by the satisfaction of the customer. Training should ensure that the technical basis of services is not a barrier to providing quality service. The values of the individuals and the organization are the basis of what the customer will perceive when dealing with the service provider.

Particular value systems are determined by a wide variety of factors, and it is not practical to suggest that an organization can be composed of completely like-minded individuals with identical personal values. Not only does the real world not work like that, it would be a remarkably dull place if it did. What is important is that the service management team have complementary skills and values that ensure that the values of the organization are displayed to the customer every time they deal with the service.

Perhaps the most fundamental of all values required in a service provider organization is honesty. The basis of any service relationship is one of trust. There is no point trying to tell users that there is nothing wrong with a service when there is, as they are more likely to be aware of the fact than the service provider. Above all else, any commitment to do something on the part of the service provider must be fulfilled or, if it is impossible, adequate notice and explanation must be provided. It is worth noting that there are also times when doing more than you say you are going to do can create problems, both in terms of inflicting the unexpected on the user and in setting an expectation for future actions.

The second core value is that the whole business should be fun. Nobody wants to deal with miserable people. If contact between service provider and service recipient is enjoyable whatever the circumstances, this is far more likely to lead to a better relationship and higher customer satisfaction. This applies equally to the service recipient in their dealing with the service provider. From my experience I can guarantee that a customer who rants and raves at the help desk is less likely to get a problem resolved than someone who is polite and reasonable. They are also less likely to have the telephone answered next time they call.

The values are the building blocks of the organization and of the customer's perception of the organization, and as such are vitally important. However, no amount of cultural brainwashing can ensure that a certain set of values are held by every individual. Thus the service management process starts when new staff are recruited. Too often, especially when looking for relatively scarce technical skills in service management staff, organizations disregard the alternative skills set that encompasses areas such as the ability of the individual to deal with people in potentially volatile situations and the value set that they bring to the organization. These "soft" skills may be more important in the long term for a service organization than any amount of technical capability.

This further emphasizes the fact that the service management team encompasses everyone in the organization, even the human resources function. Service is not just the help desk or the implementation team. The concepts and the commitment need to encompass every function and every individual. There are times when the abilities and contribution of every individual will be required to cope with the variety of events that are ordinary in the continued provision of a quality service.

Sadly, there are days when however good the organization and however good the people, things will go horribly wrong. There are days when the uninterrupted power supply will mysteriously be interrupted, when the continuous power supply will discontinue, and the only safe assumption is that every fail-safe system will fail. On top of that, every member of the help desk will have a viral infection and they will all call in sick. Of course there are other days when every target is met or exceeded, every aspect of the service has 100% availability, and customers are calling just to congratulate you on the service provided.

Both the good and the bad days are part and parcel of service provision, and both need to be managed. There is a tendency to accept the good days and manage only the bad days, but both can be valuable if managed correctly. Clearly in the case of service problems the emphasis is on resolving the problems, but there are additional elements that are sometimes overlooked.

First, admit everything. If it is a bad day and things are wrong that are affecting the service, then confess immediately. The users who know that the service is unavailable are less likely to try to use it and become dissatisfied as a result. The hope that the users either have not noticed or will never know that there is a problem is a false one. Likewise, if it is a good day, then also admit it. Ensure that people know, particularly in service provider management, that things are going well. There is no requirement to escalate only bad news.

Second, learn from the experience, both good and bad. There is normally a reason for a problem happening once, but no excuse for it happening twice. When good things happen there is no excuse for failing to ensure they happen again, by proceduralizing and documenting what was done, or just ensuring that everyone is aware of why things went well.

Finally, understand why something happened and always look for the positive. Reacting to problems with witch hunts and apportionment of blame will ensure that nothing ever progresses above the mediocrity of process management. Creativity and initiative will be hidden behind strict adherence to documented procedures. The processes should be in place to prevent things going wrong, but it is normally only action outside the process that can create the service quality that generates customer satisfaction.

The management of service provision is probably one of the most intellectually and emotionally challenging activities in business today. It is a constant battle against the odds to continually provide and enhance the quality of service that can be provided by advanced technically complex system capabilities. Service is about simplicity and about doing the simple things well: understanding the customer and their needs and objectives, responding to them in a way that supports everybody's objectives, and making it an enjoyable experience in the process. It's also about people—making people feel good about the services they use.

If you can do that, it is a formula for business success. When an organization has reached the stage of asking "Why don't our customers like dealing with us?" then it is an uphill struggle against a perception that it will be hard to change. If we could start with a clean sheet of paper and as the first question of any service ask "Why will the customers like to deal with us?" then we may be able to implement services that create customer satisfaction.

About the Author

Richard Hallows is the general manager of application services at Cable & Wireless Business Networks, London, England. Prior to this he was a service development manager for IBM Information Network. He earned his BSc in economics and politics from the University of Bristol.

Bibliography

Albrecht, Karl, and Ron Zemke, *Service America*, New York: Warner Books.

Katz, Bernard, *How to Turn Customer Service into Customer Sales*, Aldershot, Hampshire, England: Wildwood House.

Norman, Richard, *Service Management: Strategy and Leadership in Service Businesses*, New York: John Wiley and Sons.

Tack, Alfred, *Profitable Customer Care*, Stoneham, MA: Butterworth Heinemann.

Index

The Artech House Telecommunications Library

Vinton G. Cerf, Series Editor

Advanced Technology for Road Transport: IVHS and ATT, Ian Catling, editor

Advances in Computer Communications and Networking, Wesley W. Chu, editor

Advances in Computer Systems Security, Rein Turn, editor

Analysis and Synthesis of Logic Systems, Daniel Mange

Asynchronous Transfer Mode Networks: Performance Issues, Raif O. Onvural

A Bibliography of Telecommunications and Socio-Economic Development, Heather E. Hudson

Broadband: Business Services, Technologies, and Strategic Impact, David Wright

Broadband Network Analysis and Design, Daniel Minoli

Broadband Telecommunications Technology, Byeong Lee, Minho Kang, and Jonghee Lee

Cellular Radio: Analog and Digital Systems, Asha Mehrotra

Cellular Radio Systems, D. M. Balston and R. C. V. Macario, editors

Client/Server Computing: Architecture, Applications, and Distributed Systems Management, Bruce Elbert and Bobby Martyna

Codes for Error Control and Synchronization, Djimitri Wiggert

Communication Satellites in the Geostationary Orbit, Donald M. Jansky and Michel C. Jeruchim

Communications Directory, Manus Egan, editor

The Complete Guide to Buying a Telephone System, Paul Daubitz

Computer Telephone Integration, Rob Walters

The Corporate Cabling Guide, Mark W. McElroy

Corporate Networks: The Strategic Use of Telecommunications, Thomas Valovic

Current Advances in LANs, MANs, and ISDN, B. G. Kim, editor

Design and Prospects for the ISDN, G. Dicenet

Digital Cellular Radio, George Calhoun

Digital Hardware Testing: Transistor-Level Fault Modeling and Testing, Rochit Rajsuman, editor

Digital Signal Processing, Murat Kunt

Digital Switching Control Architectures, Giuseppe Fantauzzi

Digital Transmission Design and Jitter Analysis, Yoshitaka Takasaki

Distributed Multimedia Through Broadband Communications Services, Daniel Minoli and Robert Keinath

Distributed Processing Systems, Volume I, Wesley W. Chu, editor

Disaster Recovery Planning for Telecommunications, Leo A. Wrobel

Document Imaging Systems: Technology and Applications, Nathan J. Muller

EDI Security, Control, and Audit, Albert J. Marcella and Sally Chen

Enterprise Networking: Fractional T1 to SONET, Frame Relay to BISDN, Daniel Minoli

Expert Systems Applications in Integrated Network Management, E. C. Ericson, L. T. Ericson, and D. Minoli, editors

FAX: Digital Facsimile Technology and Applications, Second Edition, Dennis Bodson, Kenneth McConnell, and Richard Schaphorst

FDDI and FDDI-II: Architecture, Protocols, and Performance, Bernhard Albert and Anura P. Jayasumana

Fiber Network Service Survivability, Tsong-Ho Wu

Fiber Optics and CATV Business Strategy, Robert K. Yates *et al.*

A Guide to Fractional T1, J. E. Trulove

A Guide to the TCP/IP Protocol Suite, Floyd Wilder

Implementing EDI, Mike Hendry

Implementing X.400 and X.500: The PP and QUIPU Systems, Steve Kille

Inbound Call Centers: Design, Implementation, and Management, Robert A. Gable

Information Superhighways: The Economics of Advanced Public Communication Networks, Bruce Egan

Integrated Broadband Networks, Amit Bhargava

Integrated Services Digital Networks, Anthony M. Rutkowski

Intelcom *'94: The Outlook for Mediterranean Communications*, Stephen McClelland, editor

International Telecommunications Management, Bruce R. Elbert

International Telecommunication Standards Organizations, Andrew Macpherson

Internetworking LANs: Operation, Design, and Management, Robert Davidson and Nathan Muller

Introduction to Error-Correcting Codes, Michael Purser

Introduction to Satellite Communication, Bruce R. Elbert

Introduction to T1/T3 Networking, Regis J. (Bud) Bates

Introduction to Telecommunication Electronics, A. Michael Noll *ntroduction to Telephones and Telephone Systems,* Second Edition, A. Michael Noll

ntroduction to X.400, Cemil Betanov

he ITU in a Changing World, George A. Codding, Jr. and Anthony M. Rutkowski

itter in Digital Transmission Systems, Patrick R. Trischitta and Eve L. Varma

and-Mobile Radio System Engineering, Garry C. Hess

AN/WAN Optimization Techniques, Harrell Van Norman

ANs to WANs: Network Management in the 1990s, Nathan J. Muller and Robert P. Davidson

he Law and Regulation of International Space Communication, Harold M. White, Jr. and Rita Lauria White

ong Distance Services: A Buyer's Guide, Daniel D. Briere

leasurement of Optical Fibers and Devices, G. Cancellieri and U. Ravaioli

leteor Burst Communication, Jacob Z. Schanker

linimum Risk Strategy for Acquiring Communications Equipment and Services, Nathan J. Muller

lobile Communications in the U.S. and Europe: Regulation, Technology, and Markets, Michael Paetsch

lobile Information Systems, John Walker

larrowband Land-Mobile Radio Networks, Jean-Paul Linnartz

letworking Strategies for Information Technology, Bruce Elbert

lumerical Analysis of Linear Networks and Systems, Hermann Kremer *et al.*

ptimization of Digital Transmission Systems, K. Trondle and Gunter Soder

acket Switching Evolution from Narrowband to Broadband ISDN, M. Smouts

acket Video: Modeling and Signal Processing, Naohisa Ohta

he PP and QUIPU Implementation of X.400 and X.500, Stephen Kille

rinciples of Secure Communication Systems, Second Edition, Don J. Torrieri

rinciples of Signals and Systems: Deterministic Signals, B. Picinbono

rivate Telecommunication Networks, Bruce Elbert

adio-Relay Systems, Anton A. Huurdeman

adiodetermination Satellite Services and Standards, Martin Rothblatt

esidential Fiber Optic Networks: An Engineering and Economic Analysis, David Reed

ecure Data Networking, Michael Purser

ervice Management in Computing and Telecommunications, Richard Hallows

etting Global Telecommunication Standards: The Stakes, The Players, and The Process, Gerd Wallenstein

Smart Cards, José Manuel Otón and José Luis Zoreda

The Telecommunications Deregulation Sourcebook, Stuart N. Brotman, editor

Television Technology: Fundamentals and Future Prospects, A. Michael Noll

Telecommunications Technology Handbook, Daniel Minoli

Telephone Company and Cable Television Competition, Stuart N. Brotman

Teletraffic Technologies in ATM Networks, Hiroshi Saito

Terrestrial Digital Microwave Communciations, Ferdo Ivanek, editor

Transmission Networking: SONET and the SDH, Mike Sexton and Andy Reid

Transmission Performance of Evolving Telecommunications Networks, John Gruber and Godfrey Williams

Troposcatter Radio Links, G. Roda

UNIX Internetworking, Uday O. Pabrai

Virtual Networks: A Buyer's Guide, Daniel D. Briere

Voice Processing, Second Edition, Walt Tetschner

Voice Teletraffic System Engineering, James R. Boucher

Wireless Access and the Local Telephone Network, George Calhoun

Wireless Data Networking, Nathan J. Muller

Wireless LAN Systems, A. Santamaría and F. J. Lopez-Hernandez

Writing Disaster Recovery Plans for Telecommunications Networks and LANs, Leo
A. Wrobel

X Window System User's Guide, Uday O. Pabrai

For further information on these and other Artech House titles, contact:

Artech House
685 Canton Street
Norwood, MA 02062
617-769-9750
Fax: 617-769-6334
Telex: 951-659
email: artech@world.std.com

Artech House
Portland House, Stag Place
London SW1E 5XA England
+44 (0) 71-973-8077
Fax: +44 (0) 71-630-0166
Telex: 951-659
email: bookco@artech.demon.co.uk